Hydrogen Energy in a Sustainable Future

Phantom or Panacea?

Mark Glucina

Kozo Mayumi

CRC Press
Taylor & Francis Group
Boca Raton London New York

CRC Press is an imprint of the
Taylor & Francis Group, an **informa** business

A SCIENCE PUBLISHERS BOOK

First edition published 2026
by CRC Press
2385 NW Executive Center Drive, Suite 320, Boca Raton FL 33431

and by CRC Press
4 Park Square, Milton Park, Abingdon, Oxon, OX14 4RN

© 2026 Mark Glucina and Kozo Mayumi

CRC Press is an imprint of Taylor & Francis Group, LLC

Library of Congress Cataloging-in-Publication Data (applied for)

ISBN: 978-1-032-42147-6 (hbk)
ISBN: 978-1-032-42149-0 (pbk)
ISBN: 978-1-003-36142-8 (ebk)

DOI: 10.1201/9781003361428

Typeset in Times New Roman
by Prime Publishing Services

For my parents, Ann and Laurie
MG

Foreword

This book will be very useful for those trying to find their way in the chaotic universe of contrasting claims about the future role of hydrogen in the expected energy transition. In fact, the "hydrogen economy" has been, for decades, an iconic image of a rosy future of clean unlimited energy guaranteeing the sustainability of our economic development. However, experience in this field points to the existence of a few uncomfortable facts challenging this rosy view: (i) hydrogen is not a primary energy source (it is an energy carrier that must be produced); (ii) producing hydrogen is not easy; and (iii) as energy carrier, hydrogen is quite problematic when compared with others. What are the practical implications of these three facts? By reading this book, which is extraordinarily well organized and easy to read, one can grasp the complexity of the discussion and the diversity of factors to be considered when debating the future of hydrogen. That is, we must consider simultaneously how hydrogen is produced, transported, converted and how it is used. Unfortunately, as illustrated by the various sections of this book, this analysis is not easy in our modern industrial sectors.

The team of Mark Glucina and Kozo Mayumi represents a clear example of "renaissance scientists". They have written books and papers in physics, economics, engineering, post-normal science, bioeconomy, energy analysis, and ecological economics. In this book they managed to show an additional type of expertise, by including in their analysis, also the geopolitical and economic aspects that are shaping the research and implementation of hydrogen projects across different countries of the world. The book does not have a dry academic text style, but neither is it a journalistic type of text lacking scientific rigor. The enormous amount of quantitative information is robust and well backed up by a variety of sources.

The organization of the content is simple and effective: (i) a general introduction – what is hydrogen, how it is produced, how it can be transported, what are the most important hydrogen derivatives; (ii) what are the relevant uses of hydrogen in the economy – in industry, for heat and power, for transport; (iii) what is going on in terms of policies aimed at developing hydrogen projects around the world (including in developing countries); (iv) a section of conclusions.

In this way, the book identifies and illustrates the terms of the problem by covering a variety of aspects in a field that is changing continuously. It provides updated information while also addressing the latest political changes. What is remarkable is that the text of the book avoids the quicksand of predictions, and the temptation of indicating best solutions or roadmaps. The ultimate goal of this book is to provide a sobering view on the different issues considered, using the best information available, and this is the type of result that we really need.

Mario Giampietro
European Centre for Governance in Complexity

Preface

◇◇

The Heretics was a society formed, in opposition to conservative authority and religious beliefs, at the University of Cambridge during the early twentieth century. In February of 1923, a young polymath and future superstar biologist J.B.S. Haldane presented a multifarious paper to the society under the heading of *Science and the Future*. Amongst the many observations of science's historical impact on society and predictions for the future that make up Haldane's talk, there is a single passage concerning his speculation on what energy systems would look like once fossil fuels run out. An edited excerpt is provided below.

Ultimately, we shall have to tap those intermittent but inexhaustible sources of power, the wind and the sunlight. The problem is simply one of storing their energy in a form as convenient as coal or petrol. Even tomorrow a cheap, foolproof, and durable storage battery may be invented, which will enable us to transform the intermittent energy of the wind into continuous electric power.

Personally, I think that four hundred years hence the power question in England may be solved somewhat as follows: The country will be covered with rows of metallic windmills working electric motors which in their turn supply current at a very high voltage to great electric mains. At suitable distances, there will be great power stations where during windy weather the surplus power will be used for the electrolytic decomposition of water into oxygen and hydrogen. These gases will be liquefied, and stored in vast vacuum jacketed reservoirs, probably sunk in the ground. If these reservoirs are sufficiently large, the loss of liquid due to leakage inwards of heat will not be great. In times of calm, the gases will be recombined in explosion motors working dynamos which produce electrical energy once more, or more probably in oxidation cells. These huge reservoirs of liquified gases will enable wind energy to be stored, so that it can be expended for industry, transportation, heating and lighting, as desired. The initial costs will be very considerable, but the running expenses less than those of our present system. Among its more obvious advantages will be the fact that energy will be as cheap in one part of the country as another, so that industry will be greatly decentralised; and that no smoke or ash will be produced.

Although his timing was off by 300 years, the vision that Haldane shared 100 years ago is remarkably like that which is being proposed today. In a nutshell, we shall generate electricity from renewable sources and use it to produce hydrogen from water electrolysis. This hydrogen will then be stored, transported, and consumed as a direct substitute for fossil fuel. Countries and regions with plentiful solar and wind resources will become energy exporters to those less well endowed. Hydrogen can be burned to cook with and heat our homes, just like natural gas. We can use it to power our planes, trains, and automobiles, just like petroleum products. It can even be burned in a power plant or fed through a fuel cell to produce electricity again. No more coal required. Best of all, it produces only water vapour when burned. No more greenhouse gas. Instead of a world that feeds on polluting fossil fuels, we will have a lifeblood of clean hydrogen; a 'hydrogen economy'.

Could it be that the only reasons we have not already built this Utopia are the historically low cost of fossil fuel and, until relatively recently, ignorance of their impact on the climate? Unfortunately, no. As we shall see, there are many challenges presented by a hydrogen economy.

The use of hydrogen within the energy system is a topic both timely and critical. As we enter the second half of the 2020s, the threat of climate change has never been more apparent. According to the European Union's climate monitoring programme, every year from 2015 to 2024 ranked among the 10 warmest on record, with 2024 making headlines as the first year the global average temperature reached 1.5°C above pre-industrial levels. This threshold is widely recognized by the scientific community as a marker for potentially dangerous climate impacts. Demonstrations of this danger were felt within the borders of all the major emitters in 2024: heatwaves and drought in Europe; heavy rainfall and severe flooding in China; record breaking wildfires in North America.

As governments around the world search for effective solutions to the climate crisis, hydrogen has emerged as a promising option, attracting hundreds of billions of dollars in public investment. However, while hydrogen is sometimes presented as a clean energy panacea, it is far from a silver bullet. As an energy carrier, hydrogen has significant limitations, and understanding its role requires nuance. Given the finite resources available to combat climate change and the urgency to implement solutions, it is essential that we make intelligent decisions about which technologies to support.

Hydrogen energy, and indeed the energy transition more broadly, sits squarely in the territory of what has been called post-normal science—where facts are uncertain, values are in dispute, stakes are high, and decisions are urgent.[1] In such cases, traditional scientific methods alone are often insufficient to guide good policy. Assessing hydrogen's role in the energy transition demands not just technical analysis, but also political judgment, ethical reflection, and the inclusion of multiple perspectives.

In this book, we aim to provide an accessible, balanced, and unbiased exploration of hydrogen's role in decarbonizing global energy systems. We will explore the scientific, technological, economic, and political dimensions of hydrogen, considering both its promise and its limitations. Each chapter of this book explores a different facet of the hydrogen economy. Starting with the fundamentals, we provide an overview of hydrogen's nature and history. We then dive into the various methods

---

[1] Funtowicz, S.O. and Ravetz, J.R. 1993. Science for the post-normal age. Futures 25(7): 739–755.

of hydrogen production, the complexities of transporting it and the potential of substituting hydrogen with derivative molecules instead. We explore the case for hydrogen in various sectors such as industry, heat and power, and transport – each with its own set of opportunities and obstacles. The book is structured to guide you through this journey, from the essentials of hydrogen to the detailed challenges in its use and policy implementation around the world.

Chemistry and engineering are key elements in this story; however, we have strived to ensure that prior knowledge of these subjects by the reader is not essential for understanding. Where required, calculations will be 'back of the envelope', meaning they apply simplified assumptions and rough estimates to get a general sense of the answer. We have also avoided incorporating speculative forecasts as much as possible and focused on that which is known now. While it's true that the energy transition will unfold over several decades, and that much of this discussion focuses on the near future, we take the view that placing too much weight on projections 25 years out is unwise. Vaclav Smil's writings on the poor track record of forecasting, such as in 2003's *Energy at the Crossroads,* are recommended for those that would like to understand why.

While the scientific and technical considerations are crucial, a discussion of the hydrogen economy requires addressing broader societal, political, and economic factors. The energy transition, and Sustainability more broadly, demands the balancing of competing needs and interests. As a practical definition, Joseph Tainter[2] offered a valuable framework for thinking about sustainability by identifying four essential questions that must be answered: what should be sustained? For whom? At what cost or benefit? And for how long?

While the fourth question – "how long?" – remains the most difficult to answer, the first three are particularly relevant as we consider hydrogen's place in the energy transition. The hydrogen economy, as conceptualised by many of its proponents, is based on the idea that the wealthy countries of the 'global north' can sustain their energy intensive industries and lifestyles, which have been built on the combustion of fossil fuels, by importing clean energy from the 'global south'.

As we will see, this system would be terribly wasteful in energy terms, and from a global perspective it would be more efficient to relocate industry to where the clean energy is. However, the globally efficient solution buts up against realpolitik. The nations of the south struggle to finance the large build outs of infrastructure that are required by the energy transition without the help of the north, for whom the prospect of losing traditional industries, and their associated jobs, is politically untenable. This is increasingly true given the recent swing towards protectionism and populism in Western democracies. Just how sustainable a hydrogen economy would be depends not just on its physical characteristics, but on who benefits from it and who bears the cost.

---

[2] Tainter, J.A., Taylor, T.G., Brain, R.G. and Lobo, J. 2015. Sustainability. *In*: Scott, R. and Kosslyn, S.M. (eds.). Emerging Trends in the Social and Behavioural Sciences: An Interdisciplinary, Searchable, and Linkable Resource. Wiley.

Understanding hydrogen's role within this context requires stepping back and placing it in the larger picture of the energy transition. Hydrogen connects to nearly every aspect of the shift to renewable energy—from carbon capture and energy storage to decarbonizing transport and transforming industrial processes. As we explore hydrogen's potential, we are also examining the broader energy landscape, using hydrogen as a lens through which to understand the many interconnected challenges of creating a sustainable, low-carbon future.

Acknowledgements

We would like to express our deep gratitude to Mr. R. Prim, our CRC Press editor, for his kind invitation to write a comprehensive hydrogen book in 2021. It has taken more than three years to complete, and without his patient and consistent support, this book would not have been published. We sincerely hope that he is pleased with the result.

We have tried our best to produce a book on hydrogen's role in a sustainable energy future, that is grounded in fundamentals, comprehensive and accessible. In our view, the current hydrogen bandwagon has at times looked more like the wagon of a snake-oil salesman – veering into territory that is not fertile ground for sustainability, and potentially distracting from the task of building a cleaner energy system. Our hope is that this contribution provides a balanced and constructive perspective.

We are honoured that Professor Mario Giampietro, Emeritus Professor of the Catalan Institution for Research and Advanced Studies (ICREA) and currently working for European Centre for Governance in Complexity (ECGC), has agreed to write a foreword. Kozo Mayumi owes much intellectual debt to Prof. Giampietro with whom he has enjoyed more than 30 years of collaboration on the subject of societal and ecosystem metabolic patterns.

We are also grateful to Professor Silvio Funtowicz, Professor Joseph Tainter, and Professor Sergio Ulgiati who agreed to provide feedback on an early draft and were kind enough to provide endorsements as well. We appreciate their moral support and encouragement.

Our thanks go to Ms. Yukina Hirakuri and Ms. Wenjun Yang of the Kyoto College of Graduate Studies for Informatics for their help with figures and sourcing material, and to Russell Morgan of Nel ASA for kindly providing the electrolyzer illustration.

In addition, Mark would like to thank Laurie Winkless for early advice and encouragement, and Renee, Sofia, and Luka for being themselves.

Finally, we emphasize that all responsibility for the content of these pages remains solely with us.

Contents

PART 3. Direction: Ambitions and Actions from Around the World

PART 1

Discovery: Understanding Hydrogen's Promise - And its Problems

CHAPTER 1

The Essential Hydrogen

A Brief History

As best we can tell, it all started nearly 14 billion years ago when an unimaginable cosmic explosion sent forth the building blocks of a universe. This intensely hot material rapidly spread outward like the walls of an inflating balloon, and in so doing, began to cool. After a few hundred thousand years of expansion and cooling, the primordial material began to coalesce into the first atoms. Leading these was hydrogen, the simplest, lightest, and most abundant of all chemical elements. It's estimated that 90% of all atoms in the universe today are hydrogen. Moving forward a billion years we find that gravity has consolidated these disparate atoms into immense clouds of gas that we now call nebulae. Over time, this gas compacts down into an extremely hot and dense stellar embryo. When the temperature reaches about 15 million degrees Celsius, the hydrogen atoms may participate in nuclear fusion reactions, which create the slightly heavier element helium, and releases vast quantities of energy. A star is born.

At the end of a star's life, heavier chemical elements are formed by the nuclear reactions within it. It is this cycle of stellar life and death that produces almost all the elements in existence. Stars that go out with a bang, so to speak, fling this material out into the cosmos where it will eventually contribute to new nebulae and give birth to new solar systems. Like stars, planets form through a conglomeration of gas and space dust. We are therefore profoundly fortunate that our solar system - somewhere between 4 and 5 billion years old - is relatively young and able to incorporate the full complement of elements that the first systems did not have in their pallet. Life would not be possible without key elements like carbon, nitrogen, and oxygen. Our planet, and everything on it, was made inside dying stars. We are quite literally made of stardust, and the precursor of it all was hydrogen.

While it was undoubtedly produced by others prior, the discovery of hydrogen is generally credited to an Englishman named Henry Cavendish who lived from 1731 to 1810. Born into great wealth, Cavendish was the stereotypical gentleman scientist; working out of a home laboratory and able to devote all his attention to experimentation. Anecdotes about his life paint the picture of a man that was reclusive and painfully awkward around other people, especially women. It is said he would communicate with his female servants by handwritten notes because he couldn't

DOI: 10.1201/9781003361428-2

bear talking to them. This attribute combined with his obsessive focus on areas of interest strongly suggest that he was a case of what we now call autistic spectrum disorder. Cavendish's autism can be viewed as beneficial in that it contributed to his capacity for meticulous and precise experimentation, the most famous example being his measurement of the density of earth. This value, obtained with 18th century equipment, was within one percent of the value accepted today.

During the latter half of the 18th century, the investigation of 'airs' or what we now call gases, was a fashionable scientific pursuit. By reacting different materials with acids, Cavendish was able to produce various gases and measure their physical properties. This award-winning work included the characterisation of hydrogen, which Cavendish called 'inflammable air'. At this time many scientists believed that fire was caused by a substance called phlogiston, which was a constituent of matter that would be released into the air by burning. Cavendish found that igniting a mixture of his 'inflammable air' (hydrogen) with 'common air' (oxygen) produced water. Was inflammable air pure phlogiston, or perhaps dephlogisticated water?

This was answered by a French nobleman called Antoine Lavoisier who was born in 1731 and made his fortune in tax collecting for the royal government. He died abruptly in 1794 from the guillotine of the French Republic, having fallen foul of the revolution. In between, he contributed a great deal to science including the debunking of the phlogiston theory. He correctly interpreted water as being made up of hydrogen and oxygen, and that burning hydrogen in air amounted simply to reacting it with oxygen. In fact, it was Lavoisier that named "hydrogen", adapting the Greek words for "water" and "producer". He also named "oxygen" from the Greek for "acid producer". Lavoisier also thought all acids must contain oxygen, which is incorrect. You can't win them all.

It didn't take long before the newly christened hydrogen began popping up in novel inventions for a range of applications. French chemist Jacques Charles leveraged the fact that hydrogen is lighter than air by filling a balloon large enough to give him a birds-eye view of Paris in 1783. In the early 1800s, a French Swiss inventor named Isaac de Rivaz was tinkering with engines that could be powered by an explosive charge, rather than steam. Fuelled by a mixture of hydrogen and oxygen gas, and ignited by a spark, his design was arguably the first internal combustion engine. Not long after, in the 1820s, it was discovered that intense white light could be produced by heating 'quicklime' (calcium oxide) with a flame fuelled by hydrogen and oxygen. This invention was widely used as stage lighting through most of the 1800s until it was replaced by safer and more convenient electric lights. However, the term 'in the limelight', which was a common name for the device, is still used today to refer to someone that is the focus of public attention.

Innovation often progresses by combining multiple independent discoveries. Interestingly, the discovery of hydrogen and the composition of water occurred at around the same time the electric battery was invented by Italian Alessandro Volta in 1800. The 'voltaic pile', as it is known, allowed the production of a consistent electric current which opened the door to an important new analysis technique: electrolysis. Water was among the first substances to be examined, and the results reinforced Lavoisier's conclusion that it was composed of hydrogen and oxygen. What was surprising was that hydrogen and oxygen bubbled up at different poles. When two metal plates, each connected to a battery, are submerged into a container of water, then oxygen will be found at the positive plate and hydrogen at the negative. Similar results were obtained with a broad range of substances, which lead chemists to believe that molecules are held together by a sort of electric charge. Further adventures in electrochemistry lead to the invention of the fuel cell by Welshman William Grove around 1840. The fuel cell is basically an electrolysis cell in reverse. Where an electrolysis cell uses electrical energy to split water into its constituent parts, a fuel cell combines hydrogen and oxygen to form water and captures electrical energy that is released as a result. Grove called his first cell a 'gas voltaic battery'.

Not long after this time, and no doubt drawing inspiration from discoveries such as those just mentioned, the famous French writer Jules Verne speculated that water would replace coal as our primary source of energy in the future. In an often-quoted passage from his 1874 novel The Mysterious Island, Verne writes on the question of what we shall burn when coal runs out:

> *Water! Yes, but water decomposed into its primitive elements, and decomposed doubtless by electricity, which will then have become a powerful and manageable force. Yes, my friends, I believe that water will one day be employed as fuel, that hydrogen and oxygen which constitute it, used singly or together, will furnish an inexhaustible source of heat and light, of an intensity of which coal is not capable. Someday the coal rooms of steamers and the tenders of locomotives will, instead of coal, be stored with these two condensed gases, which will burn in the furnaces with enormous calorific power.*

At the time of writing, electric motors and lighting were still not well developed or widespread in use, which perhaps explains why Verne didn't simply suggest that we electrify everything, rather than using it to split water and burn the products.

In the decades that followed, hydrogen extended beyond fiction, finding real-world applications in important industries. For starters, commercial passenger flight can be said to have begun with the German Airship Travel Corporation (DELAG) in 1910. Using modifications of the famous hydrogen-filled Zeppelin design, DELAG carried its first passengers in the years leading up to World War 1, where upon its aircraft were promptly appropriated for military service. Following the war, the use of airships increased in popularity in both the United States and Europe. Some airships were even used to cross the Atlantic Ocean, decades before aeroplanes were capable of doing so. The market for airship travel evaporated in 1937 however, with the infamous Hindenburg disaster. While attempting to land in New Jersey, the hydrogen inside the LZ 129 Hindenburg caught fire sending the airship crashing to earth in flames and killing 35 people on board. For better or worse, this was all caught on camera and to this day informs the understanding of hydrogen for many people. It's possible to find this footage with a quick internet search, and well worth doing so.

During the 19th century, industrial scale hydrogen had been produced by blowing steam over a bed of hot coal. The product was a mixture of hydrogen and carbon monoxide that was known as 'water gas'. However, this method was made obsolete during the early 20th century by the development of a more efficient process called *steam reforming*. In the presence of a catalyst, steam reacts with a hydrocarbon fuel, usually natural gas, producing a hydrogen and carbon monoxide mixture. Additional processing steps can be added to purify the hydrogen product. This is still the method that provides most of hydrogen used today. Coincidentally, two important processes that were enabled by a cheap and plentiful hydrogen supply, were also developed around the same time.

The first process is known as hydrocracking and involves bombarding large fossil fuel molecules with hydrogen under high temperature and pressure. Its purpose is to break them up into smaller and more useful molecules such as those that make up gasoline and jet fuel. During the second World War, this process was used to make liquid fuels from coal and thick oils. In the latter half of the twentieth century, development of improved catalysts made hydrocracking more economical just as the demand for high quality aviation and automobile fuels began to ramp up. As a result, hydrocracking equipment became common in oil refineries, and still use around a quarter of the hydrogen produced today.

The second process, developed in the early 20th century, is arguably one of the most important inventions in human history. In fact, it was so good it won the Nobel Prize twice. First for Fritz Haber who invented the process and then for Carl Bosch who modified it for industrial scale a few years later. The 'Haber-Bosch' process, as it is now known, combines hydrogen gas with nitrogen removed from the air to form ammonia. While ammonia is used in many chemical industries, the majority of it is used to make synthetic fertiliser. Along with crop development, irrigation, and the mechanisation of farming, synthetic fertiliser was one of the pillars of the so-called 'Green Revolution' that produced a step jump in agricultural productivity in the latter half of the 20th century. For better or worse - because the industrialisation of farming has not been without negative consequences - farming productivity is estimated to have nearly tripled since the 1960s. Without these gains, the global population could not have increased by more than two and a half times as it has. In fact, it is estimated that a little less than half the current population is dependent on the Haber-Bosch process.[1] These people also depend by extension on the production of hydrogen, as ammonia consumes over a third of the volume produced today.

Entering the second half of the 20th century, we see hydrogen technology take a step forward thanks to the burgeoning space race. Early rockets, such as the V2 developed by Nazi Germany in World War 2, were powered by burning a hydrocarbon fuel like kerosene or methanol. However, these systems were not powerful enough to get heavy spacecraft into orbit. Liquid hydrogen had been considered an excellent propellant since the early days of rocketry: reacting liquid hydrogen with liquid oxygen would provide the most efficient propulsive force available. Even so, the technical challenges were formidable. Maintaining both components in liquid form requires extremely low temperatures. As explained in a NASA publication 'Taming Liquid Hydrogen', rocket storage tanks needed a combination of insulation and venting to prevent the hydrogen from expanding rapidly and exploding due to overheating. These heat sources include the sun, the rocket exhaust gases, and the friction generated as the rocket moves through air, with top speeds approaching 40,000 km/h. What's more, liquid hydrogen tends to embrittle the metal walls of its holding tanks, as well as leak through tiny gaps in welded joints, which again could lead to explosion. It has been claimed that mastering liquid hydrogen was one of NASA's most significant technical accomplishments, and one that gave it a decisive advantage over the Soviets in the race to the moon.[2] The liquid hydrogen and oxygen fuel system is still commonly used in rocketry today.

NASA employed another emerging hydrogen technology at this time: the fuel cell. In 1959, one hundred and twenty years after the concept was discovered, British engineer Francis Bacon demonstrated a practical fuel cell that could provide enough electricity to run a pair of hair dryers. Not long after, the 'Bacon Cell' was scaled up and trialled in the provision of power for spacecraft and satellites. For space exploration, where high costs can be absorbed and weight is of primary concern, fuel cells fed by cryogenic hydrogen and oxygen were considered preferable to batteries. Additionally, fuel cells on the Apollo missions served the dual function of powering the computers and life support systems of the spacecraft, and providing drinking water for the crew. Water of course, is the by-product of reacting hydrogen and oxygen in a fuel cell.

Space exploration had captured the public's attention in the 1960s, and in an effort to wow the public, automaker General Motors decided to produce a vehicle that runs on the same technology as spaceships. Following a brief but expensive development program, they presented the Electrovan in 1966. Credited with being the first fuel cell electric vehicle, the Electrovan was a repurposed Handi-Van, which must be in the running for the least attractive exemplar imaginable for a vehicle of the future. It ran on a scaled down version of the fuel cell that would be used in the upcoming Apollo missions. The fuel cells and cryogenic tanks took up much of the van's cargo space, making it an impractical vehicle, but in fairness it was not intended to be anything more than a demonstration of concept.

Around the same time, a South African–born chemist named John Bockris became intrigued by an idea he had encountered at an academic meeting. Based on the work of a Nazi engineer, it was suggested that piped hydrogen may be a cheaper way to transmit energy than electricity. In 1972, Bockris and fellow chemist John Appleby published a paper titled *The hydrogen economy: an ultimate economy?*[3] This is the first known appearance of the term in print, although Bockris is usually credited with coining the phrase, and the paper helped establish hydrogen not just as a fuel, but as the centrepiece of a broader energy system.

Coincidentally, the first of the 1970s oil shocks would arrive a year later, bringing with it a heightened awareness of the dangers of reliance on foreign oil, and the need for alternate sources of energy. While solutions such as further oil exploration, the production of synthetic fuels from coal, and the development of wind and solar energy received more serious attention, this was a fertile period for energy studies and hydrogen was now in the mix. Interestingly, at the time of his death forty years later, after having spent much of his career kicking around the idea of the hydrogen economy, Bockris had come to the conclusion that it would in fact be too expensive, and that a 'methanol economy' would be better. However, despite the many challenges that will be discussed in this book, hydrogen energy clearly has a 'je ne sais quoi' quality. The dream has never been extinguished completely, despite interest over the last five decades fluctuating due to a combination of technological developments, environmental science, and political priorities.

The response of industrialised countries to the 1970s oil shortages resulted in both more efficient use of energy and in a larger and more geographically diverse reserve of crude oil. By the middle of the 1980s, oil prices had dropped by nearly a third compared to the start of the decade. Air pollution from the burning of fossil fuels was a concern but 'end of pipe' technology was found to be reasonably successful at managing it. Climate change was not on the radar, and the world was quite happy to burn its coal, oil, and gas. Also, nuclear energy had not yet sustained the public relations hit of the 1986 Chernobyl disaster. Consequently, the 1980s could be considered the nadir of interest in hydrogen energy over the last half century. The same could be said for alternate energy more generally. This is not to say that no progress was made of course, simply that public funding for research and development was relatively low.

Interest in renewable energy began to slowly ramp up again during the 1990s. The price of oil was fairly stable throughout the decade, so this renewed interest was likely driven by other factors. Firstly, climate change was becoming part of the global political agenda. Starting in 1992, the United Nations' Framework Convention on Climate Change was taking place, resulting in the first global emissions reduction treaty: the Kyoto Protocol in 1997. The world was realising that it needed to wean itself off hydrocarbon energy for reasons other than price and security of supply.

Another reason was that, by the early part of the decade, researchers had successfully reduced the amount of expensive platinum catalyst required for vehicle fuel cells by a factor of ten. This was still not enough to make hydrogen cars economic, but it was a substantial step forward and it sparked renewed interest in the technology. For example, in 1993 the U.S. Department of Energy had no dedicated hydrogen budget; projects received just one to two million dollars from the broader renewables allocation. However, by 1998, hydrogen and fuel cells had secured dedicated funding: ten times what it had received just five years earlier.[4]

Hydrogen energy received a further and highly visible boost in the early part of this century when US president G.W. Bush announced his Freedom Car initiative, which along with $1.2 billion dollars in funding, would take hydrogen fuel cell cars from the laboratory to the showroom. The motivations given for developing this technology included energy independence, economic security, and reduction in both local air pollution and greenhouse gas emissions. The President also highlighted the variety of domestic energy sources that could be used to produce hydrogen: natural gas, biomass, ethanol, clean coal, and nuclear. Curiously, conventional renewable energy sources were not part of the vision. Despite many automakers having hydrogen programmes, and for reasons that will be described in more detail later in this book, the hydrogen car was not a runaway success.

As this century ticked over to its second decade, the Obama administration was of the opinion that hydrogen cars were not in fact an imminent solution and cut funding for hydrogen and fuel cells to around half of what it had been at its peak.[5]

The idea of the hydrogen economy continued to simmer however, and as we entered the 2020s it began to boil.

The Nature of Hydrogen

The hydrogen atom is as simple as atoms get. It consists of a single negatively charged electron orbiting around a single positively charged proton. The electron is about 1800 times smaller than the proton, so you could imagine this as being like a moon orbiting around a very large planet, although the earth is only 50 times bigger than the moon. Scientists prefer to think of the electron as existing in a probability cloud, rather than an object in orbit. A probability cloud can be imagined as a swarm of bees buzzing around a hive where each bee is a possible location of the electron. You never know exactly where the electron is at any given point in time, but you always know the area in which it must be located. Single hydrogen atoms are highly reactive and do not last long in the real world before finding another atom to collaborate with.

Hydrogen's most natural pure form is a gas made of two atoms joined together. Molecules made up of two atoms like this are called 'diatomic' and also include the oxygen and nitrogen molecules. Diatomic hydrogen is the smallest and lightest molecule that exists, which as you can imagine makes it difficult to pin down. Hydrogen gas is 14 times lighter than air, giving it great buoyancy, which is why it is terrific for lifting balloons and airships. This lack of weight also means that it disperses rapidly in air, quickly fleeing the scene as soon as it breaks free from containment.

Hydrogen's small size has implications for storage and transport. Being the smallest molecule around, it can slip through gaps that others can't, be it a micro-crack in a storage vessel or a loose seal on a valve. We don't yet have a precise understanding of just how leaky hydrogen systems are. Fluid dynamics theory suggests that hydrogen should leak 1.3 to 3 times faster than methane, the main component of natural gas.[6] For natural gas, leakage from the well to the customer is generally estimated at between 0.5% and 2%. While not as well studied, reported estimates for hydrogen are more or less aligned. Reported production losses span from near zero to 9%, while compressed hydrogen transport is thought to lose 0.02% to 5% in pipelines and 0.3% to 2.3% in tube trailers. The uncertainty is even greater for liquid hydrogen, with combined transport and handling losses estimated between 2% and 20%.[7]

Hydrogen is so small that it can even be absorbed by metals. The idea that a gas can move through metal may seem strange based on our everyday experience, but metals are not completely solid at a microscopic level. They are in fact made up of tiny three-dimensional grains, like how an apartment building is made up of many rooms. Hydrogen, like unwelcome mice, is small enough to slip through these rooms but tends to gather in the walls between them. These walls, known as 'grain boundaries', trap hydrogen and cause it to form bubbles that exert pressure and weaken the material. In metals, this phenomenon is called hydrogen embrittlement because the weakness brought about from the presence of hydrogen bubbles in the grain boundaries makes the metal more likely to crack or break. Steel containers that are commonly used for holding other pressurised gases may therefore be unsuitable for hydrogen. Storing and transporting hydrogen requires careful engineering and materials selection.

As the lightest substance, hydrogen also has a very low density under normal conditions. A good way to visualise density is by comparing specific volume, which is the amount of space required to hold a given weight of a substance. In Table 1.1 below, hydrogen is compared to other common gases using the units of cubic metres per kilogram (m^3/kg). The smaller the number, the less space is required to hold a kilogram of the gas. We can see that a kilogram of air at room temperature takes up 0.8 m^3, which is about the size of a standard single-door refrigerator. A kilogram of methane takes up almost twice as much space, say a large double-door refrigerator. Hydrogen on the other hand, would take up almost as much space as 15 standard refrigerators.

Table 1.1. Specific volume of selected gases.

	Hydrogen Gas	Air	Methane Gas
Specific Volume (m^3/kg)	11.9	0.8	1.5

This shows firstly why storing any gas at room temperature and pressure is impractical, and secondly why it is so much more impractical for hydrogen. For our convenience, gases are generally stored and transported either at high pressure or in a cold liquid form. The difficulty in cooling and pressurising a gas is directly related to its density, so these processes are that much more difficult for hydrogen.

When pressurising a gas, either to store in a vessel or push along a pipeline, a mechanical device is required to force the molecules to move closer together. There are many ingenious methods for achieving this but the most common takes the form of a piston squeezing the gas inside a cylinder. Molecules have a natural tendency to resist being squished together, and so they push back. As with any time we push against a force, energy is required.

A useful way to visualise the amount of energy required to compress a gaseous fuel, is to express it as a percentage of its energy content. Hydrogen is commonly stored at a pressure in the range of 350 to 700 atmospheres, and the branch of science known as thermodynamics can be used to calculate the theoretical minimum energy required to achieve this. Compressions to 350 atmospheres theoretically requires a little over 3% of the energy contained in hydrogen, while compression to 700 atmospheres uses around 3.5%. In the real world however, energy is lost because of the inefficiencies in equipment and also because gas heats up when it is compressed and this heat energy escapes into the environment. We can minimise this heat loss by doing the compression very slowly but as the saying goes, time is money, and so when a lot of hydrogen is required, the equipment has to work faster. The most efficient real-world processes will use around 5% of the energy contained in the hydrogen, but the average compression process will use closer to 10%.[8]

However, the energy cost of making high-pressure hydrogen is a bargain in comparison with the cost of making liquid hydrogen. Any gas must be cooled below its boiling point before it will become liquid. Due to our familiarity with water, we tend to think of liquid as being the natural state of a fluid, and that we need to add heat to make it a gas. However, this is because the boiling point of water is 100°C. In the case of hydrogen, the boiling point is –252.9°C. This is the second lowest boiling point of all known substances, behind only Helium's –268.9°C. To put that in context, the theoretical temperature at which a substance has no energy at all is –273.15°C. This temperature is known as 'absolute zero' and is thought to be impossible to achieve, and yet the boiling point of hydrogen is only 20°C warmer.

A great deal of ingenuity and specialised equipment is required to extract enough heat to achieve this low a temperature. The theoretical minimum energy to liquify a kilogram of hydrogen is 3.9 kWh, which is 12% of its energy content.[7] In practice, liquefaction processes use multiple cooling and compression steps, which require a lot more than the theoretical energy. Typical liquefaction plants consume a whopping 30–36% of the energy contained in the hydrogen, although there are claims that new innovative processes could halve that amount. Regardless of whether it be 12%, 36%, or something in between, making liquid hydrogen is energy intensive. This is not ideal if the hydrogen is being produced for its energy content.

While we're on the topic of hydrogen's energy content, the number most commonly used is the lower heating value (LHV): 33.3 kWh/kg. This is the amount of energy released from burning a kilogram of hydrogen, assuming that all the steam produced during combustion escapes. If the energy from that steam could also be captured, then burning hydrogen would release 39.4 kWh/kg, which is known as the high heating value (HHV). However, most real-world combustion systems can't do that, making LHV the more realistic number. We will default to this value throughout the book.

Now, the amount of energy in a kilogram of hydrogen is about two and a half times greater than those of conventional fuels. This suggests that it would also be an excellent fuel, but there is a catch. Hydrogen's very low density means that it is difficult to pack a lot of it into a tank, so its energy content on a volume basis is actually very low. Remember that 1 kg of hydrogen would take up the space of 15 refrigerators, so if used as a fuel, hydrogen must be either liquified or stored at high pressure.

Table 1.2 compares the amount of energy that is contained in a cubic metre of some common fuels. This quality is known as the 'energy density'. When discussing fuels for transportation in particular, energy density is very important because all transport modes, be it aircraft, ship, or automobile, have a limited amount of space in which to carry fuel. Because the amount of fuel determines how far a vehicle can travel, we prefer those that can pack a lot of energy into a small amount of space; those that have a high energy density.

To help visualise the relative densities, we include a row showing how many days an average American house could run on the equivalent of a cubic metre of fuel. In reality, these fuels cannot be converted into electricity without losing a significant portion of their energy content. The actual number of days would be a lot less, and depends on the efficiency of the technology used to generate electricity from fuel. This comparison is solely to demonstrate the differences in energy density.

Table 1.2. Physical properties of hydrogen and common fuels.

Fuel	State	Energy Content [LHV] (kWh/kg)	Energy Density (kWh/m³)[1]	Days Powering an American House (days/m³)[2]
H₂ (STP)	Gas, 1 atm	33.3	3	0.1 (2.3 hr)
Compressed H₂	Gas, 200 atm	33.3	507	17
Compressed H₂	Gas, 680 atm	33.3	1250	41
Methane (CNG)	Gas, 200 atm	13.9	1906	63
H₂	Liquid	33.3	2359	78
Methane (LNG)	Liquid	13.9	5811	193
Propane (LPG)	Liquid	12.7	6525	217
Gasolene	Liquid	12.4	8653	288

Source: 1. US Department of Energy. (2001).[9] 2. Assumes the average home uses 30 kWh/d.

Looking at the table, it's clear that liquids pack more punch than gases. Liquid hydrogen holds almost twice as much energy as compressed hydrogen at 680 atmospheres, and liquid natural gas (LNG) has around three times the energy of its compressed form (CNG). But hydrogen, even in its densest forms, still lags behind traditional fuels. Compressed hydrogen only carries about a quarter of the energy of CNG at the same pressure, and liquid hydrogen has just a quarter of the energy you'd get from gasoline. This means that if we want to swap fossil fuels for hydrogen, we'll need a lot more of it to do the same job.

Release of the energy stored in hydrocarbon fuels is almost always achieved by burning it. In chemical terms, burning—also known as combustion—is a reaction between a substance and oxygen, in which the products contain less energy than the original material. The excess energy is released into the environment as heat, often accompanied by light. Simplified combustion reactions for hydrogen and hydrocarbons, such as natural gas and gasolene, may be written as follows.

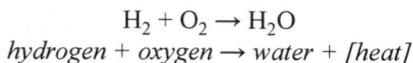

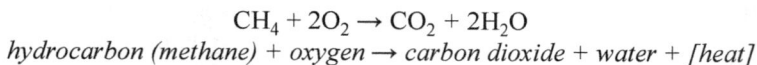

$$H_2 + O_2 \rightarrow H_2O$$
hydrogen + oxygen → water + [heat]

$$CH_4 + 2O_2 \rightarrow CO_2 + 2H_2O$$
hydrocarbon (methane) + oxygen → carbon dioxide + water + [heat]

Being an energy dense molecule, hydrogen is an excellent fuel for combustion. Hydrogen burns with a near colourless flame. In fact, a hydrogen flame is nearly invisible in daylight, which makes it a bit more dangerous than other flames we are familiar with. Hydrogen flames are a little hotter than those of common fuels too, burning at around 2200°C in standard conditions, compared to 1900°C for natural gas. Hydrogen also has a wider band of flammability than other common fuels. Figure 1.1 shows upper and lower flammability limits, which define the proportion of gas in a mixture with oxygen that will burn under standard conditions. If the percentage of the fuel gas is greater than the upper limit, there is not enough oxygen

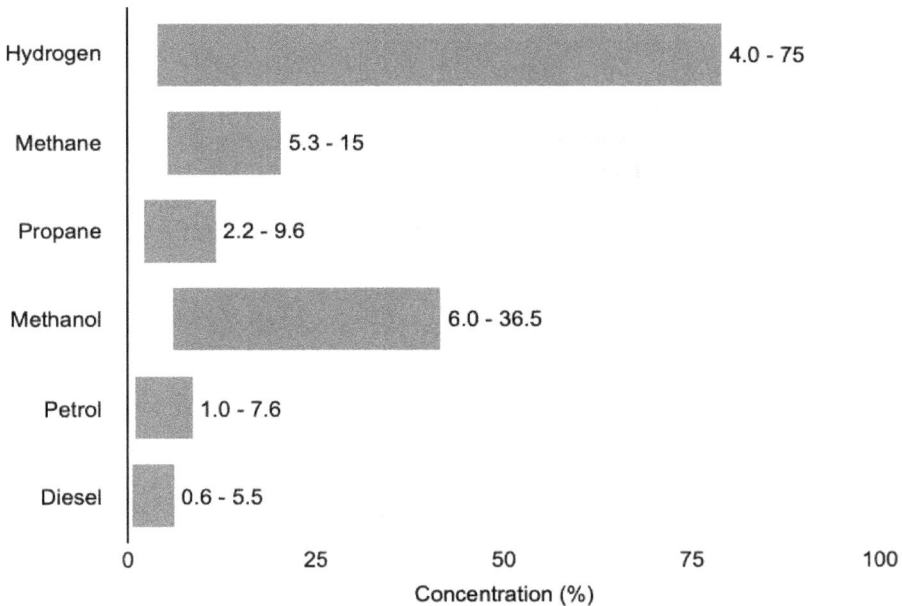

Figure 1.1. Flammability Limits for Hydrogen and Common Fuels. Source: College of the Desert. (2001). Hydrogen fuel cell engines and related technologies: A primer for students. U.S. Department of Energy.

to support combustion. If it is less than the lower limit, there is not enough fuel to burn. Note that these substances only burn in their gaseous state. In the case of liquid fuels, like methanol and gasoline, there is a layer of vapour which always exists on the surface of the liquid that does the burning. The shimmering, wavy-looking air you can sometimes see when pumping gas is the light refracting through this vapour.

From the chart, we can see that the flammability band for hydrogen (4–75%) is much wider than for any of the other substances. The amount of energy needed to ignite a hydrogen-oxygen mixture is a lot less than for the others too. An ignition source could be a flame, a spark, or even a discharge of static electricity. This means that the conditions needed to start a hydrogen fire are much easier to achieve compared to conventional fuels. Hydrogen also has low electrical conductivity, meaning that flow or agitation can lead to the buildup of static electricity—potentially generating sparks that may ignite the gas. Special attention must be paid to ensure that hydrogen equipment, such as pumps and pipelines, are electrically grounded.[9]

A wide flammability range also means a wide explosive range. Gas explosions happen when combustion is contained in a structure, like a storage tank or a building, trapping the heat and allowing pressure to build until it violently destroys the containment. This pressure buildup doesn't necessarily take a long time; it can happen in a fraction of a second. Hydrogen has about 2.5 times the energy of hydrocarbons on a weight basis, which means it also has 2.5 times the explosive force. Combine this with the wide flammability range, ease of ignition, and propensity to leak, and it's easy to imagine the substantial explosion risk that hydrogen presents.

This was famously demonstrated at the Fukushima Daiichi nuclear power plant in 2011, following a tsunami which cut off electricity supply to the nuclear reactors and disabled the cooling systems. The reactors, which were designed to generate steam by heating a continuous flow of water, instead became hydrogen generators when the steam reacted with exposed zirconium alloy used to house the nuclear fuel. In three of the plant's reactor buildings, this hydrogen eventually escaped the reaction chamber and found its way to the upper sections of the building where it joined forces with oxygen to violently remove the roof. Interested readers will be able to retrieve footage of these impressive explosions via a simple internet search.

Returning to the chemistry of combustion, hydrogen is often touted as a climate friendly fuel because it does not produce carbon dioxide when consumed, as hydrocarbon fuels do. While this is true, it is not the whole truth. Firstly, the combustion reaction that produces only water assumes that hydrogen is burning in pure oxygen. In most real-life situations, combustion occurs in air, which is nearly 80% nitrogen gas (N_2). At temperatures over 1300°C, nitrogen reacts with oxygen to form nitric oxide (NO) and nitrogen dioxide (NO_2), which are collectively referred to as 'NOx'. While not a greenhouse gas per se, NOx has long been known to contribute to local air pollution and respiratory illness. Hydrogen and hydrocarbons both burn hot enough to produce NOx and require careful engineering to minimise emissions. When hydrogen is consumed in a fuel cell however, producing electricity via an electrochemical reaction rather than combustion, then only water is produced.

The second and arguably more important point to note, is that hydrogen itself is considered an indirect greenhouse gas. This means it doesn't have a warming effect on its own, but it does interact with other compounds in the atmosphere to magnify their warming. To understand how it does this, we must introduce the hydroxyl radical (OH). This naturally occurring chemical species has been called the detergent of the atmosphere because of its ability to help wash pollutants out of the sky. It does this by reacting with certain gases to make them water soluble so they can be removed from the atmosphere by rain and snow.

One of these is methane: a strong greenhouse gas, 34 times more potent than carbon dioxide. Hydrogen is another. In effect, hydrogen in the atmosphere gobbles up some of the detergent that could be put to better use cleaning away methane. This results in a longer life for atmospheric methane, and proportionally more warming. In addition to this effect, which seems to be the most important, hydrogen can also increase the formation of other greenhouse gases like ozone and stratospheric water vapour. Taken altogether, hydrogen's global warming potential is estimated to be in the range 8.8–14.4.[10] This means that every kilogram of hydrogen released will have the same impact on global warming as around 12 kilograms of CO_2. It is therefore important that hydrogen not be allowed to leak into the atmosphere.

Well or Pail?

Energy is surprisingly difficult to define. We use the word freely in daily life, and everyone more or less knows what it *means*, though few would be confident explaining what it *is*. An almost religious notion, we know it through its works – when something moves, heats up, or illuminates. In physics, it's usually described as the capacity to do work; to drive change. Indeed the capacity to change, to endlessly transform, is the essence of energy. Yet it is always present. It cannot appear out of nowhere and it cannot vanish. Natural law prohibits this. So, whenever we access energy, we are tapping into the material world.

An energy source is the point where we place the tap. The well from which we draw. Put another way, an energy *source* is that which can be extracted or harnessed directly from the biosphere. Whether that source is renewable or not greatly influences the pressure we apply on planetary limits, and that hinges on whether the energy is captured from a *stock* or a *flow*.

Non-renewable energy sources include the fossil fuels: coal, oil, and natural gas. They are termed fossil fuels because they are the product of fossilised plants and animals – long-dead living things that have been transformed deep within the earth. These resources exist as *stocks*, which means we extract them from limited pools of material – like chocolate chips in a cookie. Fossil fuel stocks are considered finite because they take millions of years to form, but we can extract and burn them in a matter of weeks. On that timescale, they cannot be replenished.

In contrast, renewable resources implies that we are capturing *flows* of energy.

Solar power captures the flow of light from the sun. Wind power harnesses the movement of air caused by the earth's surfaces heating and cooling at different rates. Hydropower co-ops the energy of water rushing downhill under the influence of gravity. Returning the water uphill again is the work of the sun, which drives evaporation of surface water and formation of rain clouds. Sources of biomass energy are numerous; some examples being wood, algae, and even dung. Ultimately, biomass is the product of living things and is reliant on the sun to power the chemistry of life.

Even fossil fuels, which were once biomass, may be considered products of the Sun. We see a common element here. Almost all our energy, be it a stock or a flow, is ultimately sourced from the Sun.

The one exception is nuclear energy, which harnesses the immense power that glues atoms together. The radioactive element uranium is most often used as nuclear fuel. While uranium was likely produced in some long-dead star, it does not come from our sun. Nuclear energy is generally regarded as non-renewable because the uranium is a stock that must be mined from the earth. An important difference between nuclear energy and the other non-renewable energy sources is that its use does not produce greenhouse gas emissions. However, nuclear power plants do pose the risk of rendering a large swath of the region around them uninhabitable should the process get out of control as was the case with the Chernobyl nuclear plant in 1986. Nuclear energy also creates radioactive waste that must be carefully managed in perpetuity.

Now, an *energy carrier* is a substance that contains useful energy and can be easily moved around. Crucially, it can be moved from the location where the energy source is captured to the location where it will be used. Our titular pail is the energy carrier of our metaphorical well. Fossil fuels like coal, oil, and natural gas, are both energy sources and energy carriers because they are readily transportable.

Renewable energy sources, for the most part, are not. We cannot catch wind in a bag or bottle up sunlight. Instead, these flows must be transformed into something that we can transport and utilise. By far, the most convenient way of doing this is by converting them into electricity. Electricity is an extremely versatile energy carrier. It can be created from any energy source and transmitted almost instantaneously over long distances via wires, or stored in batteries for use on demand. It is also clean to use. Apart from some waste heat, electricity creates no by-products as it is consumed. And the services that electricity can provide seem limitless: lighting, heat, mechanical motion, communication, and computation. No other energy carrier can be directly consumed by such a wide range of devices.

Notably for our story, electricity can also be used to make hydrogen. Thus, any energy source that can be used to generate electricity can also be used to produce hydrogen (as long as water is also available). This makes hydrogen production almost as versatile as electricity. In principle, hydrogen can be stored and transported either as a liquid or a compressed gas. It can be burnt to produce heat or power a combustion engine. Hydrogen can also be fed through a fuel cell to generate electricity. These properties position hydrogen as a potential energy carrier.

Hydrogen is also often touted as a future 'energy source', but is this accurate? Recall that an energy source is by definition, directly available in the biosphere. It's true that there is a lot of hydrogen on earth. More precisely, there is a lot of hydrogen in the sea. Most hydrogen in the biosphere is found within water: bound up with oxygen in the iconic H_2O molecule. It is estimated that the sea is about 10% hydrogen by weight. Hydrogen is also an important ingredient in living things. It is one of the definitive elements in organic compounds; a wide-ranging class of molecules that include the building blocks of life such as proteins, fats, and carbohydrates. The remnants of living things are also rich in hydrogen. Fossil fuels belong to a class of chemical compounds known as 'hydrocarbons' – so named because they are made primarily of hydrogen and carbon atoms. In fact, their hydrogen content is a major reason why they release so much energy when burned.

All these sources of hydrogen contain additional chemical elements and require processing if we wish to have hydrogen in its pure diatomic form. Indeed, almost all the hydrogen we use today is won through the ingenious deconstruction of hydrogen-containing substances like coal, natural gas or water. This hydrogen cannot be regarded as an energy source.

For hydrogen to truly be considered an energy source, it must form naturally, but natural hydrogen is hard to come by. For every million molecules of gas in the atmosphere, less than one of them will be hydrogen. Until very recently it was also widely believed that hydrogen did not exist in meaningful quantities within the Earth's crust. It was thought that even if hydrogen was somehow present in the ground it would react with oxygen to form water, or get gobbled up by microbes, or migrate up to the surface and float away due to its small size. For a number of reasons, it seemed unlikely to hang around. People didn't really expect to find it and so largely didn't bother to look. In fact, the equipment used for gas testing often employed hydrogen as the carrier gas, which helps push the sampled gases through the equipment.[11] This effectively makes any hydrogen that may have been in the sample invisible. Despite all of this, accidental discoveries of natural hydrogen have been made.

In 1987, the village of Bourakébougou in Mali was attempting to drill a water well. Instead of water, they found a strange wind which, as the story goes, was set on fire when one of the drillers peered into the hole with a lit cigarette in his mouth. Smoking is hazardous to your health.

After attending to the burns on the unfortunate driller, the well was sealed up and largely forgotten. Twenty years later, a company exploring the region for fossil fuels realised that the strange winds were deposits of natural hydrogen gas. After further exploration, 24 bore holes around Bourakébougou were found to contain some amount of hydrogen. In 2012, a pilot project was set up whereby natural hydrogen was drawn from the ground and burned to produce electricity for Bourakébougou village. The well provides around 1500 m^3 of gas daily, which is composed of 98% hydrogen.[12] Assuming that the generator is 25% efficient, this is enough electricity to power around thirty American homes. Interestingly, it was recently reported that the pressure from this well has not decreased after eleven years of use, as would be expected from a natural gas deposit. If anything, it has slightly increased from 4.5 to 5 atmospheres of pressure. This suggests that the hydrogen is being replenished at around the same rate, or slightly faster, than it is being used. It could be a renewable resource.

Not long after the discovery in Mali was publicised, a review of naturally occurring hydrogen was published by a Ukrainian chemist named Viacheslav Zgonnik.[13] This paper catalogued instances of naturally occurring hydrogen around the world, including a large number from the former Soviet Union, which were largely unknown to western researchers.[14] Unlike in the west, Soviet engineers had been looking for hydrogen because they mistakenly believed that oil was the product of carbon reacting with hydrogen in the depths of the earth, rather than pressurised fossils. They were sampling for hydrogen in hopes of striking oil. Dr. Zgonnik states that hydrogen was frequently found in this region not because it occurs there more frequently than elsewhere, but because people were looking for it. He concluded that the common perception around naturally occurring hydrogen as of 2020 was incorrect, quoting the Wikipedia article on hydrogen fuel as an example: "pure hydrogen does not occur naturally on Earth in large quantities".

It is fair to say that this perception has changed dramatically since then. Naturally occurring hydrogen now goes by several names: white hydrogen, gold hydrogen, and geologic hydrogen. In early 2024, the U.S. Senate Committee on Energy and Natural Resources held a hearing titled *Opportunities and Challenges Associated with Developing Geologic Hydrogen in the United States*. Since then, scientists from the U.S. Geological Survey have published a prospectivity map showing the likelihood of finding geologic hydrogen across the continental United States. They also released the results of a modelling study estimating that the Earth holds between 1,000 and 10 billion megatonnes of geologic hydrogen, with a most probable value of 5.6 million megatonnes.[15]

Recall that global consumption of hydrogen currently sits around 100 megatonnes per year; 0.1% of the lower-end estimate. If the modelling is correct, geologic hydrogen represents a humongous energy source. However, it is likely that most stocks will be too deep, too remote, or too small to be economically recoverable.

Nevertheless, the hunt is on. According to the consultancy Rystad Energy, at the end of 2020 there were 10 companies prospecting for geologic hydrogen; at the start of 2024 there were 40. These firms are now searching all around the world, including Australia, Canada, the Republic of Korea, the United States, and throughout Europe. Meanwhile, President Macron of France has promised "massive spending" on geologic hydrogen, and the US has already introduced a $20 million subsidy programme for related technology.

Complicating the search is the fact that it's still not clear exactly how hydrogen gas is formed underground. One controversial hypothesis is that there is a vast well of hydrogen near the Earth's core, and that this can travel to the surface along faults and tectonic plate boundaries. Another possibility is that radiation, emitted from elements in ancient rocks, splits water molecules into gaseous hydrogen and oxygen. Perhaps the most popular hypothesis currently is a process in which certain forms of hot, iron-rich rock, chemically react with water. Essentially, the iron in the rock 'rusts' by grabbing water's oxygen atom, liberating the hydrogen. This may occur by the following reaction:

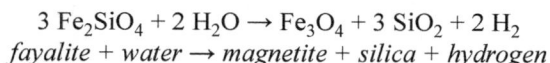

$$3 \ Fe_2SiO_4 + 2 \ H_2O \rightarrow Fe_3O_4 + 3 \ SiO_2 + 2 \ H_2$$
$$\textit{fayalite} + \textit{water} \rightarrow \textit{magnetite} + \textit{silica} + \textit{hydrogen}$$

Assuming this is correct, likely locations for hydrogen will be those with iron-rich formations of igneous or metamorphic rock. If so, there is probably very little overlap with fossil fuel resources, which are almost always found in sedimentary rock.

While no exploitable reserves have been found yet, there is buzz that the price of natural hydrogen extraction could be very low. The Bourakébougou well is said to produce hydrogen for around $0.50 per kg, which is less than half of the cost of producing hydrogen from natural gas; the lowest-cost alternative. However it's considered that this would be an outlier, with prospectors targeting costs closer to $1/kg.[16]

As to the climate impact of geologic hydrogen, this very much depends on the composition of the resource. Modelling has shown that a resource with minimal methane content would produce less than 0.4 kg CO_2/kg H_2, while a resource with a methane content closer to 20% would emit around 1.5 kg CO_2/kg H_2. Either way it would be vastly superior to hydrogen produced from natural gas, which typically emits between 10 and 13 kg CO_2/kg H_2.

As promising as all this sounds, geologic hydrogen extraction remains entirely theoretical. As will be explored in Chapter 3, transporting hydrogen is significantly more challenging than transporting fossil fuels. For geologic hydrogen to be economically viable, deposits will need to be found in large quantities, at high concentrations, and close to end users. Until this is demonstrated on a large scale, it would be wise to assume that any exploitation of naturally occurring hydrogen will be opportunistic and relatively small – like in Bourakébougou.

Feedstock or Fuel?

Almost all the hydrogen used today is made from fossil fuel, with the associated greenhouse emissions discharged to the atmosphere. With a carbon intensity of between 10–13 kg CO_2/kg H_2, this equates to around 1 billion tonnes of CO_2. For context, this is greater than the global aviation industry's emissions, which the International Energy Agency (IEA) reported as 950 Mt in 2023. Current hydrogen production is clearly unsustainable, which begs the question – what are we using it for?

Today, hydrogen is used almost entirely as a feedstock in industrial processes, where it is valued for its chemical properties – specifically, its ability to react and bond with other elements to produce useful compounds. China is by far the largest consumer, using nearly 30% of the 95 million tonnes (Mt) that were produced globally in 2022.[17] This is more than twice the amount used by either the United States or the Middle East, who at 12 Mt each, were the territories with the second-highest consumption. Europe and India were next on the leaderboard, with 8 Mt each.

Currently, most of the world's hydrogen is gobbled up by just two industries: ammonia production, where it is a building block of the NH_3 molecule, and oil refining, where it's used to clean and upgrade fuel. Together, these two industries accounted for 78% of the hydrogen produced in 2022. Relatively small amounts are also employed when making electronics, glass, and various chemicals. Hydrogen is not currently used in any meaningful way as an energy carrier. Nor is it selected as an input purely for its energy content, apart from in niche applications like rocketry. Furthermore, hydrogen is usually produced close to where it is used – often in the same chemical plant – so storage and transport are not major concerns.

In a hypothetical hydrogen economy, both of those things would change. Clean hydrogen would be produced where conditions are most favourable (where renewable energy is cheap and abundant) and then shipped long distances to be used as a fuel rather than a chemical input. So, before we can understand where hydrogen might fit in the future, we need to look at the energy system we depend on today. A system that is still overwhelmingly based on fossil fuels.

Figure 1.2 shows how the world's energy use is split across different types of energy carriers, while Table 1.3 highlights just how heavily each carrier still depends on fossil sources. The unit used here – the exajoule (EJ) – is common in global energy statistics because it keeps the numbers manageable. The prefix 'Exa' denotes a billion billions, which written out in full, is a one followed by 18 zeros. So, even though the numbers are small, the quantities of energy that they represent are massive. In 2023, the total amount of energy used by humanity (excluding food) was 445 exajoules. Note that the total energy produced was actually 642 EJ. The difference between 'production' and 'final consumption' is the universe's tax: the heat that is inevitably lost whenever energy is converted from one form to another. Overall, our current fossil fuel dominant system is losing just over 30% of the energy it procures.

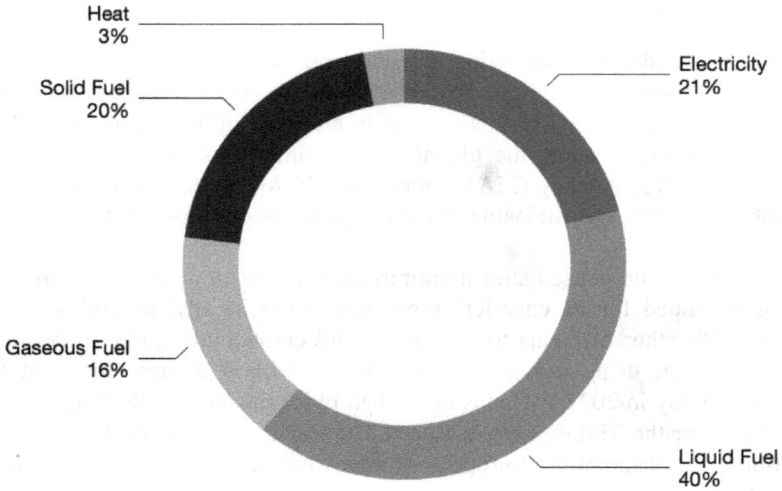

Figure 1.2. Global Energy Consumption by Energy Carrier in 2023, 445 EJ. Source: IEA (2024).[18]

Table 1.3. Fossil fuel share of primary energy by carrier in 2023.

Electricity	Gaseous Fuel	Liquid Fuel	Solid Fuel
60%	99%	97%	59%

Source: IEA (2024).[18]

At 40%, liquid fuels are the most common way that we consume energy. A small fraction of this is biofuel—made from recently living biomass—but 98% is derived from oil. These fuels are mostly burned in internal combustion engines to power transport across land, sea, and air. Gaseous fuels, which almost entirely come from natural gas, account for 16% of our energy and are used more or less equally for heating buildings and running industrial processes. Solid fuels, primarily coal, fill 20% of our energy needs and are used mostly in industry.

Altogether, around two-thirds of the energy we consume comes in the form of coal, gas, and oil products. Environmental sustainability requires this number to be much, much lower.

Unlike the three fossil fuels, the fourth major energy carrier – electricity – is not a primary energy source and must be generated. As Figure 1.3 shows, the original energy source for most of our electricity is once again fossil fuel. Around 60% of generation is based on coal, oil, and gas, while renewables and nuclear power provide the rest. Fossil-fired electricity will also have to go if we are to have a low carbon world.

So, net-zero requires that around 298 EJ of fossil fuels and 67 EJ of fossil-fired electricity be replaced with clean energy. It's no exaggeration to say that this grand energy transition is one of the most formidable tasks in human history, and if successful, will stand as one of our greatest accomplishments. However, we

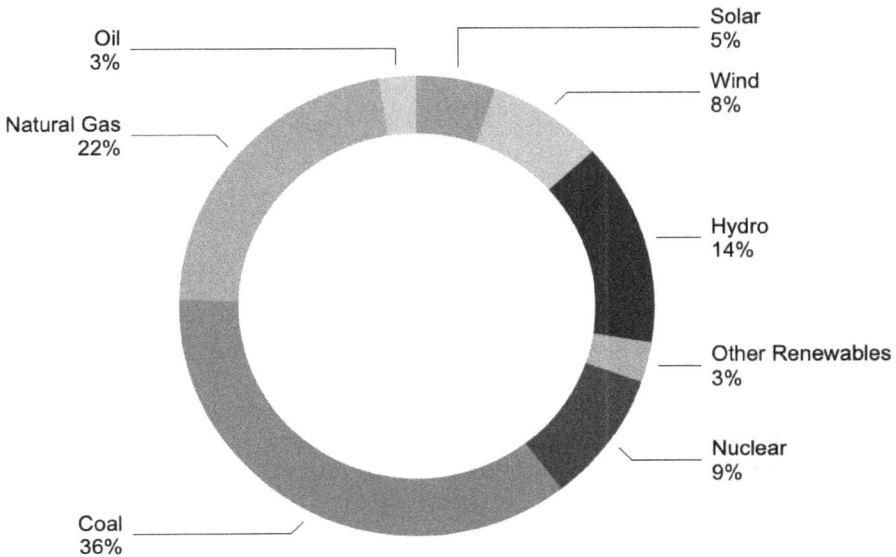

Figure 1.3. Global Electricity Generation by Primary Energy Source in 2023, 93 EJ. Source: IEA (2024).[18]

should keep in mind that global energy demand is a moving target. As the population grows and low-income regions develop, the demand for energy will increase. On top of this, emerging power-hungry technologies will add to the load significantly. The IEA estimates that data management – including both the computers that store information and the networks that move it – already consumes 2–3% of global electricity. In addition, so-called cryptocurrency mining, which involves computers racing to solve complex puzzles in exchange for digital coins, accounted for another 0.4% in 2022. The IEA estimates that their combined electricity use could double by 2026, matching the current demand of Japan.

At the same time, improvements in energy efficiency are expected to alleviate some of the pressure on our energy systems. In particular, electric vehicles and heat pumps can deliver the same services using a fraction of the energy required by their fossil-fuelled counterparts. The Swiss Army knife theory of hydrogen holds that it is also a clean replacement for fossil fuels in transport and heating. However, hydrogen technologies tend to be far less efficient than direct electrification, as we can see from Table 1.4. It's also important to note that the overall efficiency of hydrogen solutions is even lower than the values given in the table, once production, storage, and transport are considered – but we'll explore that in later chapters. The energy transition demands a laser focus on efficiency, because the less energy we need, the less infrastructure we must build—making the whole transformation faster, cheaper, and more likely to happen in time to prevent catastrophic climate change.

The high efficiency of electric technologies, combined with the fact that electricity can be produced without greenhouse gas emissions, is the one-two punch that makes direct electrification the preferred decarbonization strategy. Modelling, such as the IEA's Net-Zero Scenario, shows that if we can rapidly swap existing

Table 1.4. Indicative efficiency of common technologies.

Service	Fossil Fuel Technology	Electric Technology	Hydrogen Technology
Motive Power	Internal Combustion Engine (20–25%)	Electric Motor (85–90%)	H_2 Fuel Cell (45–50%)
Heat	Gas boiler (85–95%)	Heat Pump (300–500%)	Hydrogen boiler (85–95%)

technology with efficient electric alternatives, future energy demand may even be lower than it is now. Whether that plays out remains uncertain, but regardless of whether overall demand rises or falls, hundreds of exajoules of fossil energy will need to be replaced with renewable electricity.

Hydrogen-based energy may still have a role to play, particularly in situations where direct electrification is not feasible. However, as will be explored throughout these pages, there are a host of reasons why clean hydrogen is not a direct replacement for fossil fuel stocks, and at the risk of burying the lead, nobody is claiming that it is.

Table 1.5 lists projections of how much hydrogen will be required in 2050, assuming that the world meets its target of net-zero at this time. These are not predictions, but the results of scenario building from five different organizations. This includes the Hydrogen Council, which naturally tends to promote a rosy view of hydrogen's role in the energy transition, given that its membership is largely comprised of companies with an interest in hydrogen. Taken together, these projections suggest that for the world to reach net-zero, hydrogen need only provide something like 10–20% of our final energy consumption. To be sure, this is a lot more than the 100 Mt of hydrogen we use today for non-energy applications, but it is a long way from the 80% that fossil fuels currently provide.

We should note that these figures date back to 2021, when hydrogen enthusiasm was arguably at its peak in the current hype cycle. Most of the organizations listed update their projections annually, and subsequent versions have had much lower hydrogen requirements. Prominent clean energy research firm BloombergNEF's 2024 Net Zero Scenario lists 390 Mt of hydrogen demand in 2050. This figure

Table 1.5. Estimates of 2050 hydrogen consumption in net zero emission scenarios.

Organization	Projected Hydrogen Consumption in 2050 (Mt)	Percentage of Final Energy Consumption in 2050
IEA	520	13%
IRENA	610	12%
ETC	800	18%
BNEF	800	22%
Hydrogen Council	650	22%

Source: IRENA (2022).[19]

is nearly 25% lower than BNEF's 2023 projection and less than half of its 2021 estimate, which reflects the growing understanding of market realities during this period. Similarly, the IEA's 2023 Net Zero Roadmap places 2050 demand at 388 Mt, while IRENA's 2024 update to its 1.5°C Scenario has renewable hydrogen with a 14% share of final energy consumption, which equates to 333 Mt.

While acknowledging the considerable uncertainty in these estimates, we'll use a rounded figure of 400 Mt as a ballpark estimate for global hydrogen demand in 2050. This amount contains roughly 48 exajoules of energy, which is about 10% of today's global energy consumption. These numbers offer a sense of scale for the current vision of a future hydrogen industry. They also highlight that hydrogen is not expected to play the lead role in the energy transition. Rather, it is best viewed as a supporting actor: complementing electrification rather than stepping into the role long held by fossil fuels.

Big Numbers: Chapter 1

- **100 million tonnes** - Current global hydrogen consumption, used almost entirely for its chemical properties, rather than its energy content.

- **400 million tonnes** - Projected global hydrogen requirement in 2050 under net-zero emission scenarios. The increase reflects hydrogen's expected use as an energy carrier.

- **5.6 trillion tonnes** - Estimated global geologic hydrogen resources: a potentially enormous energy source, yet the feasibility of economic extraction remains unproven.

- **2.5x** - Hydrogen's energy content compared to fossil fuels on a weight basis, however...

- **0.25x** - Hydrogen's energy content compared to equivalent fossil fuels on a volume basis. If tank sizes are equal, hydrogen provides a quarter of the energy.

- **12x** - The global warming potential of a hydrogen compared to carbon dioxide. Hydrogen is an indirect greenhouse gas.

CHAPTER 2

Making Hydrogen

∞∞∞

The Hydrogen Colour Wheel

Hydrogen is associated with a spectrum of colours, each reflecting a distinct production method and its environmental impact. Dark, dreary colours are used for the dirtiest methods. Black hydrogen is made from coal, while grey hydrogen is sourced from natural gas. Low carbon methods receive more cheerful colours that are associated with the natural world. Blue hydrogen is still made from fossil fuel, but the associated carbon dioxide emissions are captured. The highest climate credentials are reserved for green hydrogen, created by harnessing the power of renewable energy to split water molecules through electrolysis. If the electrolysis is powered by a nuclear reactor, rather than renewables, the hydrogen produced is said to be Pink. Pyrolysis is a novel method that thermally dismantles methane, leaving behind hydrogen and solid carbon. It can be fed by natural gas while producing no direct emissions. So, sitting somewhere between blue and green, it has been given the fitting moniker of turquoise hydrogen. Geologic hydrogen, collected directly from the earth, also goes by the aliases white and gold. Less commonly, hydrogen that is produced from a waste stream is called orange. These naming conventions are summarized in Figure 2.1.

None of these names are officially standardized. In fact, different colours are sometimes used for the same method, which can create confusion. For example, both Black and Brown have been used for coal derived hydrogen, sometimes to distinguish the type of coal. Some also refer to electrolytic hydrogen from solar electricity as yellow, which, given that solar is a renewable source, could equally be labelled green. Regardless, the colours are used informally in policy and research contexts for their convenience.

However, convenient as it may be, and cute for marketing purposes, the colour system is inadequate as a description of climate impact. In can even be misleading. As will be explained in more detail throughout this chapter, the precise carbon footprint of a hydrogen process depends on how well it is run and can vary significantly. For this reason, there have been many attempts at defining variations of 'clean' or 'low carbon' hydrogen based on measurement of the CO_2 released during the production process. There may actually be too many. In its *2024 Global Hydrogen Review*, the International Energy Agency lists more than 50 certification systems

DOI: 10.1201/9781003361428-3

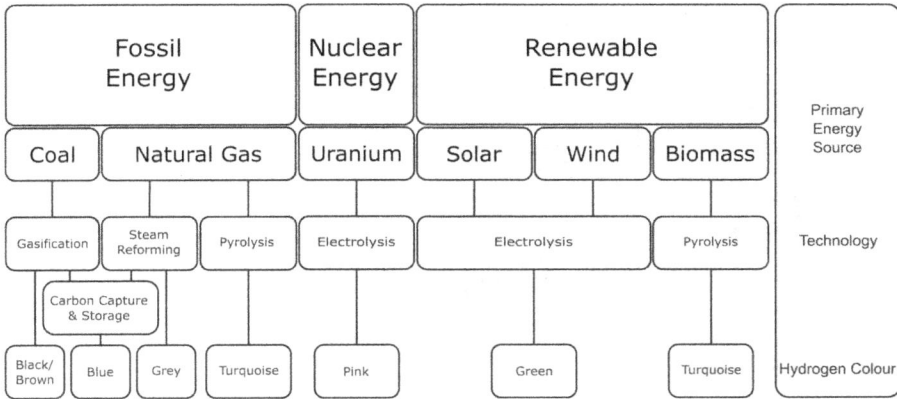

Figure 2.1. Hydrogen production routes.

Table 2.1. Current hydrogen production primary energy sources.

Natural Gas	Coal	Refining by-product	Low-emission methods
62% (59 Mt)	19% (20 Mt)	16% (15 Mt)	0.7% (0.65 Mt)

Source: IEA (2023).[1]

and frameworks spread around the globe. This presents problems for hypothetical trade in clean hydrogen, as one nation's green may be another's greenwash. We are still a long way from this being a major problem, however. Low-emission hydrogen production is miniscule.

As Table 2.1 shows, natural gas is the most common hydrogen feedstock by far, accounting for 62% of production. Second is coal at 21%, with most of the remaining hydrogen generated as a byproduct of oil refining. Hydrogen from low emissions sources constituted less than 1% of the total production and almost all of this was again made from natural gas, but with the carbon dioxide captured. We must keep in mind however that capturing the carbon dioxide does not imply it is kept out of the atmosphere. Currently, less than 10% of the CO_2 captured industrially is placed in permanent storage.[1] The rest is used by industry, so each specific case would need to be examined carefully to determine how low carbon it truly is. Only a very small amount of hydrogen – less than 100,000 tonnes – was produced from the electrolysis of water.

Fossil-Fuelled Hydrogen

Steam Methane Reformation (SMR)

The most common method used for large scale hydrogen production is steam methane reformation. This has been used for over a century and, as the name suggests, involves the use of steam to strip hydrogen atoms away from methane molecules. Methane (CH_4) is the main component of natural gas. A simplified diagram of the SMR process is provided as Figure 2.2.

1:gas purification unit 2:reformer 3:waste heat boiler

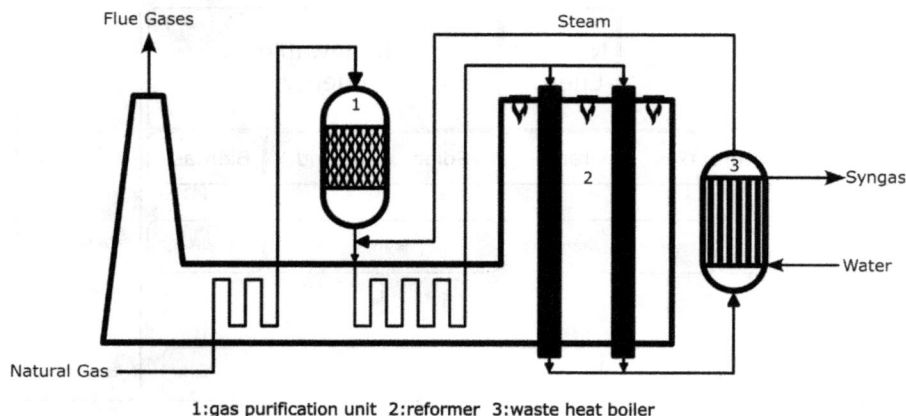

Figure 2.2. Simplified SMR process.

In 2021, SMR used around 200 billion cubic metres of natural gas, which was 6% of global consumption.[2] Natural gas has historically been very cheap to acquire via extraction from underground stocks. Consequently, the price of hydrogen produced via SMR has been the lowest available. In 2021, this price ranged between 1 and 2.5 US dollars per kilogram of hydrogen.[3] The Russian invasion of Ukraine did send world gas prices soaring through much of 2022, however they have since normalised, albeit higher than pre-war levels.

In the SMR process, natural gas plays the dual role of supplying the hydrogen atoms and the heat energy that makes the process go. First, natural gas mixes with steam and a catalyst in a reactor vessel. Processes are set up differently to suit local conditions, but the reactor temperature is usually between 700–1000°C with a pressure between 3 and 25 atmospheres. In this highly energetic environment, the steam and methane molecules rearrange, or *reform*, according to the following reaction:

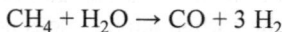

$$CH_4 + H_2O \rightarrow CO + 3\ H_2$$

methane + water → carbon monoxide + hydrogen

In a follow up reaction, the carbon monoxide product is mixed with extra steam to produce even more hydrogen. This is known as the water-gas shift reaction, and proceeds as follows:

$$CO + H_2O \rightarrow CO_2 + H_2$$

carbon monoxide + water → carbon dioxide + water

Both reactions are 'endothermic', meaning they require additional heat energy to proceed. This heat would usually be provided by burning some of the natural gas feed. Overall, the SMR process is somewhere between 70–85% efficient, meaning that the hydrogen product contains 70–85% of the energy that went into the process in the form of natural gas and electricity.[4]

The environmental burden of SMR is notable. The steam feed requires at least seven litres of water for every kilogram of hydrogen produced. Using the best technology available, the SMR process also releases around nine kilograms of carbon dioxide for every kilogram of hydrogen. When we also include gases that escape during the collection and transporting of methane, total emissions are in the 10–13 kg CO_2/kg H_2 range. This is untenable in a low carbon world. However, SMR's low-cost relative to alternatives means that it continues to be attractive. Consequently, pairing SMR with carbon capture and storage technology to produce blue hydrogen is frequently proposed as a solution for low-cost and low-carbon hydrogen. More on that later.

Autothermal Reforming (ATR)

If SMR is like cooking with a pot on the stove, ATR is a pressure cooker: cleaner, more contained, and a tick more high-tech. Both technologies produce hydrogen through the reformation of methane, but the ATR method has a touch more finesse. Its key difference is the introduction of oxygen into the reaction vessel, as shown in Figure 2.3. Oxygen's role is to combust a portion of the methane, generating heat internally. This step, known as 'partial oxidation', proceeds as follows:

$$2\,CH_4 + O_2 \rightarrow 2\,CO + 4\,H_2$$

methane + oxygen → carbon monoxide + hydrogen

If the proportions of oxygen, steam, and methane are carefully balanced, the heat released by partial oxidation can supply all the energy required for both the reforming and water-gas shift reactions. This allows ATR to maintain a stable temperature without the need for external furnaces, which are essential in SMR.

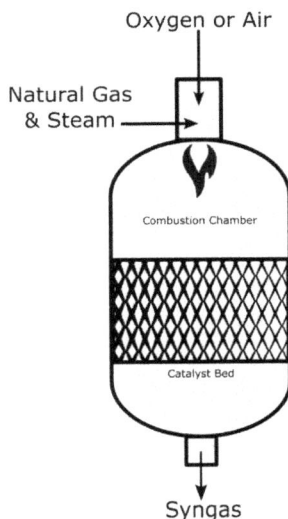

Figure 2.3. Simplified ATR reactor.

However, the savings from avoiding furnaces are more than offset by the need to build and operate cryogenic air separation units to supply pure oxygen. ATR also operates at higher pressures and involves more complex safety considerations. As a result, it generally costs more to run than its cousin and may be slightly less energy efficient overall.

For blue hydrogen production, however, ATR offers a major advantage. All the gas leaving the process is contained in a single, high-pressure, CO_2-rich stream— making carbon capture relatively straightforward and efficient. In contrast, SMR produces CO_2 in both the process gas and the furnace exhaust, the latter being at low pressure and low concentration, which is difficult and costly to capture. Because of this easier integration with carbon capture systems, ATR is often the preferred technology for new blue hydrogen projects.

Coal Gasification (CG)

Coal gasification plants are relatively expensive to build, costing perhaps three times as much as a comparable SMR plant.[5] At 60–75%, they are also less efficient at converting energy inputs into hydrogen.[4] However, gasification can still provide hydrogen at a cost similar to SMR in locations where coal is cheap and plentiful. China uses this method extensively for hydrogen production.

The role of coal in CG is slightly different from that of natural gas in SMR. Like natural gas, coal acts as a fuel, supplying heat energy to drive the reactions. However, methane brings a lot of hydrogen to the dance: four H atoms for every C atom. Coal does not. While it does contain around 5% hydrogen, that contribution is minor. Coal's actual role is to break apart water molecules, which is where most of the hydrogen in coal gasification originates. CG is even more thirsty than SMR, requiring at least nine litres of water for every kilogram of hydrogen.

A simplified explanation of the process goes like this. The coal is first dried and pulverised. This is important for speeding up the reactions because thousands of small particles have a much greater surface area than a single big lump of coal, and the surface is where reactions happen. The crushed coal is then fed to a reactor vessel which runs at a temperature in excess of 900°C. Here it is blasted with alternating jets of air and steam. The air provides a little combustion, which generates heat. The steam then reacts with the hot coal, which is mostly carbon, as follows:

$$C + H_2O \rightarrow CO + H_2$$

carbon + water → carbon monoxide + hydrogen

However, coal is not pure carbon; it is closer to 50%. In addition to hydrogen, coal contains elements like oxygen, nitrogen, and sulphur. As you can imagine, the chemistry inside a gasifier is a little more complex than the simple reaction shown above. Several unwanted gases and impurities must be removed before we have high purity hydrogen. This includes a whopping 20 kilograms of carbon dioxide for every kilogram of hydrogen; twice the amount generated from SMR. However, a possible silver lining is that the CG process lends itself to carbon capture more easily than SMR because of the way it can separate out the carbon dioxide into a concentrated stream. Consequently, additional CCS equipment may increase the cost of building a CG plant by only a few percent. In the case of SMR, adding CCS may increase the cost of the plant by as much as 70%.[5]

Methane Pyrolysis

Pyrolysis is the breakdown of organic molecules using heat in the absence of oxygen. The name gives it away: from Ancient Greek, *pyr* meaning 'fire' and *lysis* meaning 'to loosen' or 'break apart'. Without oxygen, combustion can't occur—but if enough heat is applied, the chemical bonds within the molecules break down. For example, methane can be pyrolyzed to produce hydrogen and solid carbon, as shown in the reaction below:

$$CH_4 \rightarrow C + 2 H_2$$

methane gas $\rightarrow$ solid carbon + hydrogen gas

An obvious benefit of this method is that no planet warming carbon dioxide is produced. The solid carbon, if it is of sufficient quality, can be used as a raw material for products like tires and plastics. Usually, this material would be made from fossil fuel, so there are potential bonus emissions savings here as well. The tricky part of pyrolysis is that methane molecules have strong, stable bonds, so a lot of energy is required to break them. If using heat alone, a temperature greater than 1000°C is required to get a useful rate of production.[6] This temperature can be lowered using a catalyst, but unfortunately the solid carbon tends to smother it over time. This reduces the catalyst effectiveness and requires that it is cleaned or replaced, increasing the cost and process complexity.

Pyrolysis technology has been in development for a long time and there are now several different proprietary processes. However, it is still quite new on a commercial scale. The first demonstration plants have only started to pop up in recent years. Most are producing, or expecting to produce, hydrogen on the order of hundreds or thousands of tonnes per year. In comparison, a good sized SMR facility would produce somewhere between 50,000–100,000 tonnes per year, if not more. With that said, the Monolith plant in the US state of Nebraska, which is one of the world's largest, has been producing 5000 t/y of hydrogen since 2020 and is in the process of expanding to 50,000 t/yr. This plant uses a reactor that heats methane to 1700°C with a plasma torch, as illustrated in Figure 2.4.[7] The torch is powered by renewable electricity, which it consumes at a rate of 25 kWh/kg H_2.[8] This is around half the electricity used by electrolytic production of hydrogen, although this excludes the energy content of the methane. If we include this, then the total energy input is probably closer to 90 kWh/kg H_2, which is getting close to twice that of electrolysis.

Proponents of turquoise hydrogen might say that it combines the low cost of SMR with the low carbon footprint of renewable power electrolysis. For example, Monolith's website claims that their pyrolysis reactor produces around 0.45 kg CO_2/kg H_2. However, the size of the carbon footprint depends very much on where the methane comes from. If natural gas is used, then leakage is a problem. The IEA estimates that the greenhouse emissions from methane leaks between the well and

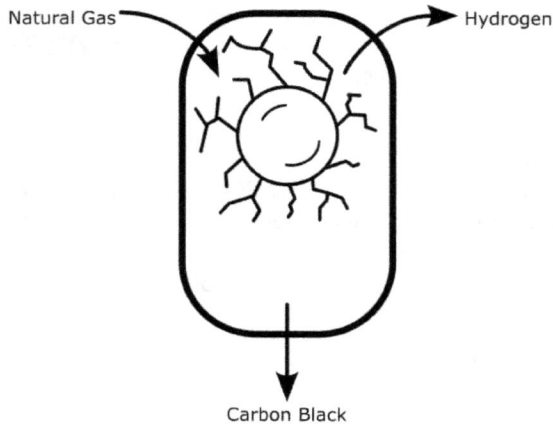

Figure 2.4. Simplified plasma pyrolysis reactor.

the end user are in the range of 4.5 to 28 kg of CO_2 equivalent per gigajoule of natural gas.[1] This translates to 0.9–6 kg CO_2/kg H_2 made from pyrolysis. The leakier end of the range is hardly low carbon.

On the other hand, If the methane is produced from a renewable resource such as crop residue or animal waste, then the resulting 'bio-methane' could possibly have negative emissions. This is because the biological feedstock captures carbon dioxide from the air while it is growing. In theory this is a wonderful concept, however it's questionable how plug and play it is. Potential feedstocks of bio-methane, and the processes for converting them, are numerous and diverse so the feasibility of any given project really is case dependent. The IEA reports that the cost of bio-methane ranges from \$7–70 per MWh, depending on how it is made. At the lower end, it is competitive with the cheapest natural gas, which costs around \$10/MWh in the US. The IEA also estimates that the sustainable global potential of bio-methane is currently about 700 TWh per year.[9] This is less than the 820 TWh of natural gas used to make hydrogen in 2022. So, there simply wouldn't be enough bio-methane available to supplant the fossil version.

In addition to price and scale, there is one other dark and dirty consideration that constrains turquoise hydrogen: every kilogram of hydrogen produced by pyrolysis is accompanied by nearly four kilograms of carbon. That's a lot of carbon. Indeed, it could be argued that pyrolysis is a carbon manufacturing process with hydrogen as a byproduct. While it's true that carbon is a useful material that is required in many different products, it is the scale of this market that gives us pause. Global demand for carbon black was around 15 million tonnes in 2023. If all this demand was met through methane pyrolysis, the associated hydrogen would only be around 4 million tonnes – 4% of current global hydrogen demand. So clearly, if we want to make a lot of hydrogen from pyrolysis, we will need to find some new uses for the associated carbon.

Turquoise hydrogen is commercial now and has attractive attributes that will likely see it play a role in the future. However, the limitations described above show why its scale will be limited, and why it's generally not given the same weight in the hydrogen energy discourse as Green and Blue.

Carbon Capture Blues

As we've just seen, the incumbent methods of making hydrogen are ingenious—but they churn out a lot of carbon dioxide in the process. The prospect of capturing carbon dioxide that is generated by a fossil-fuelled power plant or industrial process, and storing it underground, is often presented as an innovative new weapon in the war against climate change. Thus, carbon capture and storage (CCS) is also a prominent character in the story of hydrogen.

Hydrogen could continue to be made by steam methane reformation, which is the lowest cost method we have, and by squirrelling away the pesky carbon dioxide byproduct we could conceptually produce hydrogen with low cost and a small carbon footprint. Immediately, this seems an attractive solution as it would allow us to keep using the plant that has already been built, and the cheap fuels that they feed on. Simply bolt some tech onto the end of the smokestack and we can carry on with business as usual. Terrific!

In actuality, carbon capture is a technology that has been around for fifty years. It was developed in the early 1970s by the US oil industry to prolong the life of oil fields. They discovered that by injecting large quantities of carbon dioxide into wells that were almost spent, they could push out the last drops of oil. This practice is called 'enhanced oil recovery' (EOR). It's not cheap to set up and run, but neither are oil fields. A little added investment to extend their life can be economically worthwhile, especially when oil companies can obtain government subsidies aimed at keeping down fuel prices. Between 1972 and 2014, almost all the carbon capture projects in the world were to be found in the US and the purpose of these projects was enhanced oil recovery.[10]

In the last 25 years or so, carbon capture and storage has increasingly been promoted as a climate fix. Nowadays, CCS means the capture of carbon dioxide with the explicit purpose of trapping it underground in suitable geological structures, thus preventing its contribution to global warming. Enhanced oil recovery has been rebranded as CCUS; carbon capture use and storage. While the carbon dioxide used in EOR is theoretically trapped underground, the fact that in doing so it produces more polluting fossil fuel makes it a tough sell as a form of climate action. Regardless, carbon capture today is still very much focused on oil recovery. Reuters has reported that 30 of the 42 projects currently operating around the world use the captured CO_2 for EOR while 12 send it underground without boosting oil production. Indeed, 75% of the 300 million tons of carbon dioxide that had been captured up to 2022 was used for EOR.[11]

The focus of carbon capture does appear to be shifting, however. According to the Global CCS institute, 20 of the 26 projects currently in construction are being built for dedicated storage, with only six planning EOR. However, despite 50 years of development, there is still great uncertainty around how effective these projects will be, assuming they are ever switched on at all. This is because CCS is a risky undertaking with a long history of underperformance and a project failure rate greater than 80%.[12] In this context, *failure* means a project was cancelled or abandoned before completion.

A 2022 study found that just 13 major projects account for over half of the world's current operational capacity and are responsible for around two thirds of all the carbon dioxide that has been captured thus far.[11] Of the 11 for which data was also publicly available, only three have performed close to their design capacity. Six have underperformed by average of 38% and two are not currently operating. It needs to be mentioned that carbon capture plants are not designed to grab 100% of emissions either. It is said that 90% is the typical capture target but two of the newer plants – Gorgon in Western Australia and Quest in Canada – had targets of 80% and 35% respectively. Quest is performing to design while Gorgon has underperformed by 50% meaning they are both still venting something like 65% of the carbon dioxide to the atmosphere. Corporate communications intended for the public inevitably focus on how many tonnes of carbon dioxide have been captured, rather than capture rates. Always look on the bright side of life!

Capture inefficiency aside, what makes CCS such a dodgy proposition that only around a fifth of projects even get off the ground? In engineering terms, capturing the carbon is simple enough. Most commonly, carbon dioxide is captured by directing the gas stream through a reaction vessel where it contacts a liquid solvent that absorbs the CO_2. In a subsequent step, the carbon dioxide is released by changing the temperature or pressure of the solution, which then loops back around to collect more gas. You've probably seen video of an electromagnet suspended from a crane being used to sort metal in a scrap yard. Carbon capture is essentially a chemical analogue of that.

Once captured, the carbon dioxide must be dehydrated and compressed. Dehydration is required because any water left in the gas stream will react with carbon dioxide to form acid that eats metal tanks and pipework. This is usually done in a similar way to CO_2 capture; by passing the gas stream through a material that will stick to water but not carbon dioxide. Compression is used to reduce the volume of carbon dioxide, making pipeline or ship transport more cost-effective. Transport is required because carbon dioxide cannot be injected just anywhere. Storage requires particular geological features, which are not necessarily in the same location as where the carbon dioxide is produced.

All these process steps employ mature technology that has been used commercially for a long time.[13] However, 'mature' does not mean low cost and here is the first reason CCS is difficult to implement: carbon capture plant is expensive and for it to work well, it needs to be tailored to the overall process.

This is not like buying a bucket off the shelf to catch water from a leaky tap. Each project requires its own engineering design to account for the size and location of the plant, and the characteristics of the gas stream being treated. Table 2.2 shows indicative costs of capturing carbon dioxide from industrial processes. Compression, transport, and storage typically add another $20–25 per tonne under favourable conditions, although it could be double that amount for difficult cases.[5, 13]

Table 2.2. Carbon capture characteristics of important industrial processes.

Process	Typical CO_2 fraction in Gas Stream[1]	Cost of CO_2 Capture (\$/tonne)[2]
Natural Gas Processing	Over 90%	15–20
Ammonia (Haber-Bosch)	10–60%	25–35
Hydrogen (Steam Methane Reforming)	10%–20%	50–80
Iron and Steelmaking	3%–35% depending on process	40–100
Power Generation	4% for Natural Gas, 13% for Coal	50–100
Cement	18%	60–120
Direct Air Capture	0.04%	600–1000[3]

Sources: 1. Global CCS Institute (2021).[14] 2. IEA (2021).[15] 3. IEA (2023).[16]

We can see from the table that the capture cost largely follows the concentration of carbon dioxide in the gas. The lower the concentration of CO_2 in the gas stream, the harder it is to capture, the bigger the capture equipment needs to be, and the more expensive it is. This is the reason why sucking carbon dioxide out of the air, which is known as direct air capture (DCA), is not currently a feasible solution to climate change. Sadly, due to the low concentration of carbon dioxide in air, it is eye-wateringly expensive with current technology – tree planting notwithstanding.

To put these numbers into context, the price of emitting carbon dioxide due to taxes or emission trading schemes is frequently a little under \$50 per tonne.[17] Of course this varies greatly around the world from \$155/t in Uruguay to nothing in many countries. So, depending on where in the world a plant is operating, it may very well be cheaper to vent the carbon dioxide and pay the tax than to run an expensive CCS operation. In the past, plant operators that have CCS equipment in good working order have even chosen to turn it off when they were unable to sell the carbon dioxide to oil producers for EOR.[18] Put simply, CCS is purely a cost and if businesses can't find someone to pay for it, be it a CO_2 customer or the taxpayer, they are disinclined to do it.

Arguably the most successful examples of commercial CCS without EOR are to be found in Norway. In 1996, the Sleipner CCS project became the world's first commercial operation with dedicated geological storage. Snøhvit joined it in 2007. There are two key reasons why these have been relatively successful. The first is that they are both attached to natural gas processing, which as we can see from Table 2.2, produces a concentrated carbon dioxide stream and has the lowest carbon capture cost of any industrial process. In fact, natural gas processing requires that the carbon dioxide be separated as a matter of course. This is because liquifying natural gas for transport entails cooling it to less than $-160°C$, but carbon dioxide freezes at a cosy $-78.5°C$. Any carbon dioxide mixed in with the natural gas becomes solid and damages the liquefaction equipment, so it must be removed.

The second reason is that Norway has had a high carbon tax for a long time. In 1996, when Sleipner began operations, it was over $66/t and is now more than $90/t.[14] Both projects have capture capacities of a little under one million tonnes of carbon dioxide per year, so assuming a total cost around $40/t, CCS should produce savings of tens of millions of dollars per year for each project.

Proponents say that CCS would be widely adopted if governments could only get the price of carbon pollution high enough, like in Norway. They cite economics, or lack thereof, as the principal reason why CCS is not widely applied. While this has no doubt been an important factor in the cancellation of projects, there is another concern that casts doubt on its potential to address climate change in a big way – the *S* in CCS is a problem.

While the carbon capture and transport steps do use mature technologies, the storage side is much less established. Carbon dioxide is stored by injecting it vertically into the ground at a depth of at least 800 m.[13] This entails compression to around 100 atmospheres, at which point carbon dioxide becomes supercritical: a state of matter that is dense like a liquid, but flows like a gas. The idea is to inject the CO_2 into a layer of porous rock, which is situated below a layer of solid rock. Carbon dioxide can flow through the porous layer, but not the solid layer, which blocks it from percolating back up to the earth's surface. The deep porous formations most used for storage are either depleted oil and gas fields or saline aquifers—underground reservoirs filled with salt water.

Alas, as is often the case, the practice is not as simple as the theory. Despite the best efforts of well-funded experts, equipped with the most advanced technology, experience has shown that the behaviour of geological formations injected with pressurised carbon dioxide is hard to predict and can change over time.

For example, a CCS project in Algeria was halted after seven years of operation when it was found that injecting carbon dioxide had increased the pressure in the porous storage layer, causing the ground to swell like a bee sting. This movement led to cracking in the layers of cap rock that were supposed to contain the carbon dioxide, which would potentially create pathways for it to leak back up to the surface if allowed to continue. The swelling even raised the surface of the earth near the injection site by around 25 mm, causing cracking in nearby buildings.[19] The Gorgon plant in Western Australia that was mentioned previously has also faced unexpected difficulties with pressure buildup. In this case, they have elected to pump seawater out of the storage layer, freeing space in the rock for the carbon dioxide to go in. This of course adds a tidy sum to the overall project cost.

Not even the model Norwegian projects have escaped storage drama.[16] At Sleipner, it was expected that the carbon dioxide would take many years to migrate up from the injection point to the cap rock. Surprisingly, in less than three years, a large quantity of carbon dioxide was found within a layer close to the surface that had not even been identified by previous geological investigations. The gas is currently spreading out in a horizontal direction, and nobody is sure how much is there and how much farther it can go.

At Snøhvit, the rock formation that was originally selected for storage following extensive investigations, was believed to have enough capacity to last for 18 years of injection. It lasted 18 months. Unexpected pressure rise forced the abandonment of the original formation, which was replaced by a smaller and shallower stopgap, to buy time for the identification and development of a new one. This added hundreds of millions of dollars in additional cost to the project.

Given this experience, it surely requires some highly motivated reasoning to find comfort in CCS as a widespread solution to climate change, especially given that some of the current proposals are for storage fields ten times larger than Sleipner and Snøhvit. Geological surveying and modelling can only provide estimates of what lies deep in the ground, and carbon dioxide can do unexpected things, even after years of behaving as expected. Clearly, continuous monitoring is required to ensure there is no leakage or dangerous geological changes, not only while storage operations are running but also long after they have been retired. The ability to take rapid action, should it be required, also implies access to expertise, equipment, and funds, long after the operation has ceased proving income to its owners. It is fair to ask how realistic any of this is.

Running CCS equipment is also energy intensive. In the case of electricity generation, CCS ends up eating a lot of the power that is being produced: 13–18% for natural gas-fired plants and 21–44% for coal-fired plants.[20] Given that the cost of renewable energy has now dropped to the point where it is often cheaper than fossil fuels, CCS appears a poor investment by comparison. The IEA reported in 2020 that the median cost of electricity is in the $40–45/MWh range for hydro, wind, solar, and nuclear sources.[21] In contrast, the median cost of electricity from coal and natural gas is around $70–80/MWh without CCS, and $80–90/MWh with it. Renewable power is half the price of the CCS option, and it's only getting cheaper.

Now, some suggest that CCS may still have niche applications in industries other than power generation, for which low-carbon alternatives may not be feasible. Blue hydrogen, made from methane reformation with CCS, is an example of such a niche. With government subsidies on offer to help offset the price of production, companies operating in locations with cheap natural gas such as the US and Middle East may yet be prepared to invest in expensive CCS equipment. The green credentials of hydrogen made in this way would then depend on how much natural gas leaks to the atmosphere on the journey from the well to the production plant, how much carbon dioxide evades the capture system at the plant, how much hydrogen leaks from the plant, and how effective the carbon dioxide storage is.

Consequently, it is crucial that we exercise the utmost care in assessing emissions: applying a Sherlock Holmes level of attention to detail. We must be awake to the fact that assessment methods are not completely objective and can be tuned to deliver a result favourable to whoever is executing it.

An instructive case study is a widely cited 2021 report by the Hydrogen Council, which is a coalition of multinational companies – including all the oil majors – with the purpose of promoting hydrogen as an energy transition solution. This report, titled *Hydrogen Decarbonisation Pathways* assessed the lifecycle emissions for a series of hydrogen production methods and use cases. According to the methodology employed, all pathways provided emissions reductions of at least 75% compared to fossil fuels.[22]

However, scientists from the US non-profit Environmental Defence Fund (EDF) found that this study employed overly optimistic numbers. These included assuming very low leakage rates for methane (0.2–0.5%), a very high rate of carbon capture (98%), and not considering hydrogen leakage at all. They also used the global warming potential (GWP) over 100 years as the key measure. This is not unusual, as it is a common metric in these sorts of studies, but when the key gas is methane, the results are more favourable than the alternative of GWP over 20 years. This is because the atmospheric lifetime of methane, like hydrogen, is less than 20 years so its effect appears smaller with a longer time horizon.

The EDF scientists reran the calculations using what they considered to be more realistic numbers and published the results. With a methane leakage rate ranging from 0.6 to 2.1%, a carbon capture rate of 60%, and hydrogen leakage of 1 to 10%, they found that blue hydrogen pathways ranged from 70% better to 50% worse than using fossil fuels.[23] The Hydrogen Council's results were consistent with the EDF's best-case scenarios, but gave no hint of the possibility that using blue hydrogen is worse for the climate than the fossil fuels they are displacing.

A similar conclusion was reached by influential climate researchers Robert Howarth (Cornell University) and Mark Jacobson (Stanford University) in their widely discussed 2021 paper How green is blue hydrogen?[24] Focusing on hydrogen production rather than its use, they found that blue hydrogen offers only modest emissions savings compared to grey hydrogen. This is largely because carbon capture systems require additional energy—typically supplied by burning more natural gas—which significantly increases total gas consumption. Assuming that the rate of methane leakage on its journey from the well to the plant remains the same, using more gas simply means more of it escapes. In their analysis, these additional 'fugitive emissions' on route to the plant erase much of the benefit gained from capturing CO_2 at the plant.

They calculated that with a 1.5% methane leakage rate, well to gate emissions from blue hydrogen production are only 18–25% lower than those of grey hydrogen. Their best-case estimate for blue hydrogen came to 7.5 kg CO_2/kg H_2, which is far from low carbon. Assuming renewable electricity is used to power the capture equipment helps somewhat, reducing emissions to 6.25 kg CO_2 per kg H_2, but this still fails to meet any reasonable definition of "clean." This shows just how easily upstream methane leaks can wipe out the benefits of capturing carbon later on.

Howarth and Jacobsen also note that the oil and gas industry has largely been behind the promotion of blue hydrogen and suggest that this is economically beneficial from their perspective because more natural gas is needed to make blue than grey. They conclude that there is no way that blue hydrogen can be considered 'green', and that it is "best viewed as a distraction; something that may delay needed action to truly decarbonize the global energy economy".

It is fair to say that this sentiment is frequently encountered in writing from environmentalists familiar with the hydrogen space. As described previously, there are many good reasons for this:

- the high monetary and energy cost of carbon capture
- modest carbon capture rates still allow significant emissions

- our experience of carbon storage being plagued with difficulties
- scepticism that carbon stores will be well managed for centuries to come
- seemingly unavoidable upstream methane leakage eroding any gains from carbon capture
- the prospect of cheaper renewable-based alternatives

As a result, some see CCS as nothing more than greenwash; a fairy tale spun by 'Big Oil' to obscure the need to phase out fossil fuels and help them continue with business as usual for as long as possible. From this point of view, the argument that it is a bridge technology until renewable alternatives can be scaled up sounds like just another stall tactic to lock us into continued fossil fuel use. After all, process plants are expensive and built to run for decades.

If they can secure buyers willing to pay a premium, and convince governments to subsidise them in the process, gas companies will happily produce blue hydrogen. Of this there is little doubt. While the environmental concerns surrounding blue hydrogen remain significant, its continued development is likely driven by strong industry lobbying and the policies of countries like Japan and Korea, which appear more inclined to embrace it. As a result, blue hydrogen may gain traction, even if its sustainability remains suspect. It will undoubtedly remain contested ground between environmentalists advocating for cleaner alternatives and industry interests pushing for its viability.

Hydrogen from Water

Dating back to 18th century investigations of gases, electrolysis of water is the oldest known method for producing hydrogen. However, its high cost relative to fossil fuel-based methods has historically limited its use to the laboratory. That is, until climate change concerns necessitated the replacement of fossil fuels with clean energy sources. In a low carbon future, it is thought, most of our hydrogen will need to be produced via electrolysis and powered by renewable electricity. Hydrogen produced in this manner is widely known as 'green hydrogen'.

Conceptually, the electrochemical reactor or 'cell', is constructed a bit like a sandwich, as shown in Figure 2.5. Two electrodes act as the bread with some sort of electrolyte as the filling. The electrodes carry electricity into, and out of, the reaction chamber and are connected to a power source via electrical wiring. The electrodes are also where the water-splitting chemical reactions occur. One of the electrodes, known as the anode, has a positive electrical charge and produces oxygen gas. The other, known as the cathode, has a negative charge and produces hydrogen gas. The electrolyte is a substance that fulfils the dual role of allowing water and electricity to flow. Sticking with the sandwich analogy, there are different combinations of bread and filling available. The market is currently dominated by two recipes: alkaline and PEM.

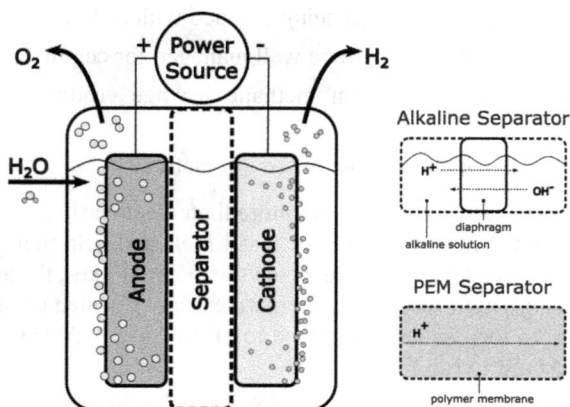

Figure 2.5. Simplified hydrogen electrolysis cell.

Alkaline Electrolysis

The alkaline electrolyzer (Figure 2.6) is the more commercially mature technology. It uses relatively inexpensive stainless steel or nickel-alloys for electrodes, and a lye solution for the electrolyte. Lye is drain cleaner: a water-based solution containing either potassium hydroxide (KOH) or sodium hydroxide (NaOH). It is used in electrolysis cells because these chemicals dissolve into charged particles that are good at carrying electrical charge through water. Specifically, it is the negatively charged hydroxide ion (OH⁻) that moves between the electrodes, completing the electrical circuit. A diaphragm, made from a special porous material that lets only negatively charged particles pass, is situated between the electrodes. This plays the role of keeping hydrogen and oxygen gas separated, which is important for both product purity, and for preventing the buildup of explosive mixtures. These gases are formed at their respective electrodes via the following reactions:

Alkaline Cathode Reaction (negative electrode)

$$4\,H_2O + 4\,e^- \rightarrow 2\,H_2 + 4\,OH^-$$

water + [electricity] → hydrogen + hydroxide ions

Alkaline Anode Reaction (positive electrode)

$$4\,OH^- \rightarrow 2\,H_2O + O_2 + 4\,e^-$$

hydroxide ions → water + oxygen + [electricity]

We can see that the amount of electrical charge generated at the anode and consumed at the cathode is the same. These reactions must happen simultaneously for the electrical circuit to sustain. For this reason, they can be considered half-reactions, with the overall reaction shown below. Note that there is twice as much water going in than there is coming out. Water and electrical energy are the two things that must be continuously fed to keep the process running.

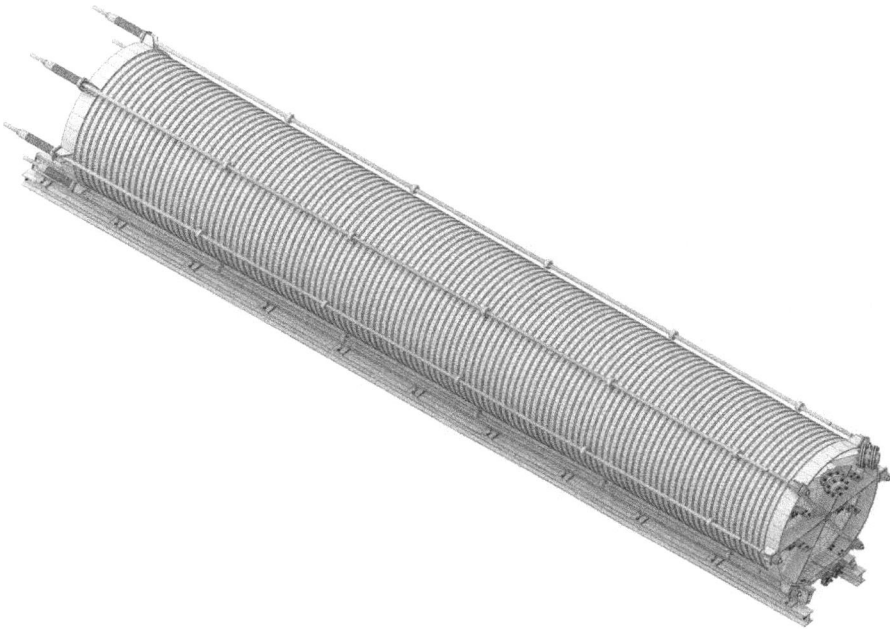

Figure 2.6. Alkaline electrolyzer stack. Provided by Nel ASA.

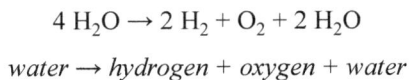

$$4 \ H_2O \rightarrow 2 \ H_2 + O_2 + 2 \ H_2O$$

water → hydrogen + oxygen + water

A variation on the alkaline cell replaces the diaphragm with a solid membrane, similar to the PEM cells below. This design, known as 'anion exchange membrane' (AEM) would operate in dilute alkaline solutions rather than concentrated lye. This is kinder on the equipment and is expected to reduce system costs. AEM electrolysers are not yet available commercially, however.

Proton Exchange Membrane Electrolysis (PEM)

The PEM, or proton exchange membrane electrolyser, is a newer technology that does the same job as its alkaline cousin, but in a slightly different way. Recall that the alkaline cell uses the negatively charged hydroxide ion as the charge carrier. PEM goes the other way by using the positively charged hydrogen ion (H^+). The hydrogen ion is simply a proton, hence the name *proton exchange*. The half reactions for PEM electrolysis are as follows:

PEM Anode Reaction (positive electrode)

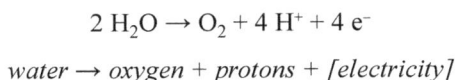

$$2 \ H_2O \rightarrow O_2 + 4 \ H^+ + 4 \ e^-$$

water → oxygen + protons + [electricity]

PEM Cathode Reaction (negative electrode)

$$4\,H^+ + 4\,e^- \rightarrow 2\,H_2$$

protons + [electricity] → hydrogen

One way of introducing protons as the charge carrier would be to use an acidic solution as the electrolyte because having a lot of free protons is what makes something acidic by definition. However, this is hard on the process equipment as highly acidic solutions tend to eat away at materials over time. PEM electrolysers get around this problem by using a solid membrane electrolyte: an ingenious sandwich filling that sits between the electrodes and allows only protons to move through. The electrodes are also different, requiring rare and expensive metals like platinum and iridium.

Currently, both PEM and alkaline technologies are being selected for new hydrogen projects. They each have relative advantages over the other, and the choice seems to depend on the specific requirements of a given project. PEM electrolysers are slightly more efficient. They convert between 67–84% of the supplied electrical energy into hydrogen, compared to 62–82% for alkaline.[4] This does not look like a significant difference but when electricity makes up 80% of the cost of producing hydrogen, a small efficiency gain can have a big effect on profitability.[2] However, PEM is currently more expensive to build due to the exotic electrode materials that are required. PEM electrolysers may cost twice as much to buy as alkaline. On the other hand, PEM tends to be smaller, lighter, and easier to handle than alkaline reactors, which weigh somewhere between 50–100 tonnes. Alkaline reactors require the lye electrolyte to be changed every 3 to 5 years. This can be a significant logistical challenge when a single electrolyser holds many tonnes of hazardous liquid, and a plant runs many electrolysers. PEM electrolysers need no electrolyte change as the solid membrane lasts as long as the rest of the equipment.

The relative ease of transporting equipment and the lack of refills can be an advantage for PEM in projects that are geographically isolated and difficult to get to. As you can imagine, this is not uncommon given that proximity to a large sunny or windy location is a major influence in selecting where to build a hydrogen plant. These locations are not always convenient.

Solid Oxide Electrolysis (SOE)

A third and relatively new electrolysis design, known as solid oxide electrolysis, has only recently started being employed in industrial settings. The scale is still small: around four megawatts at most. PEM and alkaline projects commonly plan on capacities of hundreds or even thousands of megawatts.

SOE incorporates relatively inexpensive ceramic or metal-ceramic composite materials for all components of the sandwich. The main advantages of SOE are high purity hydrogen and high energy efficiency. The latter is achieved by using a combination of thermal and electrical energy to split water. In contrast to the low-temperature alkaline and PEM machines, SOE operates at 500–1000°C and consumes steam rather than liquid water.

Because the heat energy in the reactor does some of the heavy lifting, only 37.5 kWh of electricity is required to make a kilogram of hydrogen. This is less than the theoretical room temperature requirement of 39.4 kWh/kg for water splitting, and much less than the practical energy consumption of alkaline and PEM, which is generally around 50 kWh/kg.

SOE's electrode reactions are shown below. Note that steam is fed to the cathode, and it is the oxygen ion (O^{2-}) that carries charge between the electrodes.

SOE Cathode Reaction (negative electrode)

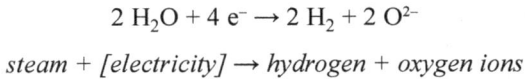

$$2 H_2O + 4 e^- \rightarrow 2 H_2 + 2 O^{2-}$$

steam + [electricity] → hydrogen + oxygen ions

SOE Anode Reaction (positive electrode)

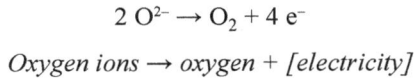

$$2 O^{2-} \rightarrow O_2 + 4 e^-$$

Oxygen ions → oxygen + [electricity]

The ceramic materials used in the SOE cell are well suited to withstand the high temperature and humidity of this reaction environment. However, as anyone who has taken a ceramic dish out of a hot oven and placed it into cold water can attest, they do not cope well with rapid changes in temperature. For this reason, they need to operate continuously and with a consistent supply of high-temperature heat and electricity. This is problematic for variable solar and wind energy, which cannot provide a constant supply on their own. Nuclear reactors however provide both constant heat and power, making SOE a strong candidate for nuclear-derived hydrogen. Alternatively, SOE could be paired with a hydrogen-hungry process that also creates waste heat. Some of these, such as the ammonia and synthetic fuel processes, will be discussed in later chapters. SOE may not be suitable in all cases, but it could be an excellent option in the right situation.

Electrolysis Challenges

Even though electrolysis is widely touted as the future of hydrogen production, it has not quite demonstrated that it is compatible with a variable power supply. At least not in a manner that produces low-cost green hydrogen. Renewable energy sources like solar and wind do fluctuate, so if an electrolyser is to be run solely on these sources it needs to be able to quickly ramp up and down its production and operate over a wide range of input current. In short, an electrolyser needs to keep pace with the ups and downs of renewable power.

Connecting an electrolyser to reliable grid power may not solve the problem either, given that most countries use fossil fuels to generate some portion of their electricity. Governments sometimes require that green hydrogen producers match their production to periods when renewable generation is occurring to maintain climate integrity, which is of course the whole reason for using hydrogen as an energy carrier in the first place.

Green credentials aside, it is also highly desirable to avoid power fluctuation because we don't want electrolysers to blow up. Keep in mind that they are built with sandwich structures, and that the respective hydrogen and oxygen producing electrodes are physically separated by a membrane or diaphragm. In theory, this separation prevents the oxygen and hydrogen mixing, but in reality, it isn't perfect.

With its small size, some hydrogen inevitably finds its way into the oxygen stream. Because the amount of oxygen produced is more or less proportional to the amount of power being fed in, this isn't a problem if the electrolyser is being supplied by enough power to produce a steady stream of oxygen. However, when the power supply drops to a low level, only a small amount of oxygen is formed and any hydrogen that sneaks through the membrane is now present in a higher concentration. You may remember from the previous chapter that just four percent hydrogen is enough to start a fire, and because engineers are very conservative with issues of safety, electrolysers are generally designed to shut down before the hydrogen concentration gets to half of this.

This safety consideration narrows the range of incoming power under which an electrolyser can operate. It's a bit like trying to ride a bicycle next to a jogger. The cyclist, which in this analogy is the electrolyser, will be fine if the jogger is keeping a good pace. However, as the jogger (or electricity supply) slows down it becomes increasingly difficult for the cyclist to maintain balance and at some point, they risk falling over and must put their foot on the ground.

This is not a deal breaker for green hydrogen in engineering terms, but it does make the economics a lot more questionable. At the time of writing, the largest green hydrogen plant in the world is located in Kuqa, Western China. The facility began operations in mid-2023 and boasts 260 MW of electrolyser capacity, powered by a dedicated solar farm and supplemented by wind from the local grid. However, reporting from the specialist news outlet *Hydrogen Insight* suggests the plant is running at less than a third of its maximum output.

This is because the electrolysers were expected to operate safely whenever power supply reached at least 30% of maximum load. In other words, the machines had an operating range of 30–100%. In practice, the minimum safe operating point appears to be closer to 50%. Given that the plant uses equipment from three different manufacturers, and all seem to share this limitation, it looks like a fundamental issue with the alkaline design.

It is also a costly one. Electrolysers are expensive to build, so the less they're used, the more each kilogram of hydrogen ends up costing. If a system can't run safely when power is low, it sits idle more often—and that drives up the price of the hydrogen it does manage to produce.

A final consideration for electrolytic hydrogen, and one that is often overlooked, is the need for fresh water. The UN's 2024 *World Water Development Report* states that:

"As of 2022, roughly half of the world's population experienced severe water scarcity for at least part of the year, while one quarter faced extremely high levels of water stress, using over 80% of their annual renewable freshwater supply. Climate change is projected to increase the frequency and severity of these phenomena, with acute risks for social stability."[25]

Theoretically, electrolysis requires nine litres of high purity water to produce one kilogram of hydrogen, which is about 30% more than is used for making grey hydrogen by SMR. If all the hydrogen we used today was green, the water consumed to make it would fill around 400 Olympic swimming pools. While that is certainly a lot of H_2O, it would be only 0.0001% of global water use by industry. The water requirement is therefore not a deal breaker for green hydrogen, but it is a consideration for project placement.

Regions that get a lot of sun, making them good candidates for solar power, often also experience water stress for the same reason. Some assumed green hydrogen powerhouses that fall into this category include Australia; Spain; North Africa and the Middle East. The solution in many instances will be to locate hydrogen production near the coast and use purified seawater. These days this is most commonly done via 'reverse osmosis', which is a mature technology that uses high pressure to force water through a super-fine membrane. The membrane acts like an extremely precise filter – trapping contaminants, minerals, and salt molecules – and letting only clean water pass through.

Adding this to a green hydrogen process obviously increases the overall cost and energy consumption, but not unreasonably so. Reverse osmosis uses around 0.0025–0.004 kWh/L of salt water. If we allow for a little waste and assume a green hydrogen plant is using 10 L/kg H_2, the water treatment is adding no more than 0.04 kWh/kg H_2 to the 50 kWh/kg H_2 required by electrolysis. This is less than a tenth of one percent. According to the industry body Hydrogen Europe, this increases the cost of a kilogram of hydrogen by less than one cent.

It's important to note that reverse osmosis also produces a waste stream of highly concentrated saltwater that can harm ecosystems if not carefully managed. The idealised green hydrogen plant would have access to a supply of fresh water, so it would not have to deal with this along with the additional costs. However, given the potential conflict between green hydrogen production ambitions and water scarcity in many regions, reverse osmosis does represent a robust technological solution.

Fringe Tech

There are a good number of additional methods that can be used to produce hydrogen, but for one reason or another are not suitable for large scale production with our current technology, and perhaps never will be. A few of the more prominent are described below.

Biological Conversion is an umbrella term for the use of specialised microorganisms that react with biomass, producing hydrogen as a waste product. These processes are also known as 'fermentation', the most well-known example of which is brewing. In beer-making, the enzymes contained in yeast break down sugars to make alcohol plus the carbon dioxide that gives beer its fizz. Similarly, certain algae and bacteria contain enzymes that can break down organic molecules and produce hydrogen gas. An advantage of biological processes is that they happen at room temperature and so do not require expensive materials and energy inputs. The main disadvantage, and the reason why they are unlikely to be used on an industrial scale, is that the rate of hydrogen production is slow and so we would need a very large number of reactors to produce meaningful quantities of hydrogen.

A second method that uses biomass as the hydrogen source is gasification. Biomass gasification works much the same way as coal gasification which was described earlier. The key difference being instead of coal, the raw material can be something like vineyard prunings, rice husk, specially grown algae, or even banana peels. Like coal, biomass gasification requires high temperatures: 700–1200°C. However, it is substantially less efficient, converting only 50% or so of the input energy into hydrogen.[4] Biomass is also often a lot more complex than coal in terms of the variety of molecules it contains. This leads to unwanted residues that can block up the equipment or contaminate the product. Biomass gasification is explicitly included in India's definition of green hydrogen and may be trialled commercially soon.

The third hydrogen production concept, and one which has received a hearty serving of academic interest, is known by the umbrella term 'thermochemical cycles'. And a jumbo-sized umbrella it is. It's estimated that somewhere between two and three thousand unique thermochemical cycles have been studied.[26]

These processes can be likened to chemical Rube Goldberg machines. Water is fed in at one end and undergoes a series of chemical reactions designed to produce hydrogen and oxygen while recycling all other reactants. A good example is the Sulphur-Iodine cycle, which is one of the more widely studied and well-regarded variants. In the first step, water reacts with sulphur dioxide (SO_2) and iodine (I_2) to form sulphuric acid (H_2SO_4) and hydroiodic acid (HI), which are then separated. Both acids are then subject to high temperature decomposition via two separate reactions. The sulphuric acid decomposes into oxygen, which is removed from the process, and a sulphur dioxide and water mixture that is recycled to the first reactor. The hydroiodic acid decomposes to the hydrogen product and iodine, which is also recycled.

Reaction 1 (Acid formation):

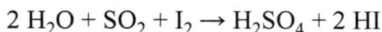

$$2\ H_2O + SO_2 + I_2 \rightarrow H_2SO_4 + 2\ HI$$

water + sulphur dioxide + iodine → sulphuric acid + hydroiodic acid

Reaction 2 (Hydroiodic acid decomposition):

$$2\ HI \rightarrow H_2 + I_2$$

hydroiodic acid → hydrogen + iodine

Reaction 3 (Sulphuric acid decomposition):

$$2\ H_2SO_4 \rightarrow 2\ H_2O + 2\ SO_2 + O_2$$

sulphuric acid → water + sulphur dioxide + oxygen

Thermochemical cycles such as the Sulphur-Iodine route are a good example of processes that work well in a laboratory but are impractical or overly expensive on an industrial scale. This sort of process requires carefully controlled conditions to maintain the desired reaction rates, and for the Sulphur-Iodine cycle specifically, this means temperatures greater than 800°C for the decomposition reactions.

That said, Japan is working towards testing this concept by pairing the sulphur-iodine cycle with a new generation nuclear reactor that is also being developed by the Japan Atomic Energy Agency (JAEA). High-temperature gas-cooled nuclear reactors (HTGR) are said to be very safe because they are designed in a way that makes a meltdown impossible. The downside of the design is that they are quite inefficient and generate a lot of waste heat. However, this waste heat could potentially be used to run the sulphur-iodine cycle, enabling a plant that produces both electricity and clean hydrogen. Handling strong acid at high temperature will add a further layer of complication, but the JAEA hopes to trial the concept within a few years and have a demonstration plant up and running by 2030.

Production Cost Comparison

Were it not for the 'inconvenient truth', as coined by Al Gore, we would happily continue to make hydrogen from natural gas in perpetuity. It's mature, low-cost, and energy efficient relative to the other options. Alas, greenhouse gas emissions must be stopped. There are options to do so, but they must be viewed in the context of how their costs compare with the alternatives in terms of both money and emissions. Alternatives that are too expensive simply won't happen. Alternatives that do not provide meaningful emissions reductions simply should not happen.

When talking about money, we often use the levelized cost of hydrogen (LOHC) as the measure of affordability. This is broadly what hydrogen costs to make when you consider the cost to build the process equipment, the cost to run the equipment, and the cost of the primary energy used to make the hydrogen. The relative importance of these different costs changes according to the production method.

For example, plants that can run continuously tend to have an edge over operations that need to stop and start. This is because firstly, they are making more product each year, but also because the building cost can be spread out over a greater number of hours. The sneakers that you wear everyday are always going to end up better value for money than the dress shoes you break out once a year.

Another important factor in LOHC is the interest payable on loans taken to fund the construction of the plant. Companies generally can't afford to fund the construction of billion-dollar plants from their cash reserves and need to borrow money to build them. When the money is this big, shifts in interest rates can dramatically impact the price of a project—and, by extension, the LOHC.

Table 2.3. Representative hydrogen production costs.

H$_2$ Production Method	Hydrogen Colour	Monetary Cost ($/kg H$_2$)	Climate Cost (kg CO$_2$e/kg H$_2$)
Steam Methane Reformation	Grey	0.7–5.7	10–13 (includes upstream)
SMR with CCS	Blue	1.2–6.9	7.5–10.5[4]
Coal Gasification	Black	1.8–3.6	19–23[5]
CG with CCS	Blue	2.0–4.7	6–9.5[5]
Methane Pyrolysis	Turquoise	2.75–3.4[2]	~ 0 (direct) 0.9–6.0 (upstream) < 0 (biomethane)
Water Electrolysis, renewable electricity	Green	3.6–12+	~ 0 (direct)
Water Electrolysis, nuclear electricity	Pink	~ 70% of Green H$_2$[3]	0.1–0.3

Sources: IEA (2024).[28] Exceptions: 2. Sanchez-Bastardo et al. (2021).[29] 3. Lazard (2024).[30] 4. IEEFA (2022), assumes 40% capture[31] 5. Liu et al. (2022)[32]

Money isn't the only cost that matters, however. When it comes to climate impact, we often compare greenhouse emissions using the measure of CO$_2$-equivalent. This is because there are many gases that have a warming effect, but the magnitude of this effect and their lifespan in the atmosphere vary considerably. For example, methane lives in the atmosphere for a mere 12 years, compared to approximately 100 years for carbon dioxide. However, methane is much more effective at absorbing solar energy, so over the course of 20 years a tonne of methane has a global warming impact similar to 86 tonnes of carbon dioxide.[27] Translating different gases into CO$_2$-equivalents allows us to compare apples and oranges. Table 2.3 presents indicative costs, both financial and environmental, for today's fossil-based hydrogen production methods and the most commonly proposed low-carbon alternatives.

The first thing to note is that none of the alternatives are as cheap as the current methods. That shouldn't be surprising as low cost is precisely the reason why SMR is used so heavily, along with coal gasification in the case of China. Regardless of how it's made, the price of hydrogen depends heavily on the cost of energy, so it does vary by location. That said, grey hydrogen typically sits at the low end of the cost scale, around $1 to $1.50 per kilogram. Blue hydrogen generally costs two and a half times more, while green hydrogen is often around five times more expensive. With a gulf this wide, clean hydrogen isn't just behind in the race—it's running in a different event. This cost cap isn't just a nuisance, it's the main obstacle to a large-scale clean hydrogen industry.

Now, if we look at the options from a climate-first perspective, then electrolysis is the clear winner. This is the reason why most of the discourse around hydrogen energy is focused on electrolysis and ways to get the costs down. In the IEA's reporting, green hydrogen starts at around $3.6/kg and is capped at $12/kg for comparison purposes. In reality, it can exceed that upper bound in some regions. The low end of the cost range can be achieved only with the highest-quality renewable resources and the least expensive equipment. In an unremarkable location, green hydrogen currently costs somewhere between $5 and $8 per kilogram. Pairing electrolysis with nuclear power could offer slightly cheaper hydrogen. Analysis by the financial advisory firm Lazard suggests that pink hydrogen is about 30% less expensive than green hydrogen in the United States.

The main reason pink hydrogen is cheaper is capacity factor. Electrolyzers can run far more consistently under a steady nuclear supply than under variable renewables. However, building new nuclear plants is much slower—and far more expensive—than installing wind or solar. Public resistance to new nuclear facilities also remains strong in many countries. Pink hydrogen could certainly play a role, but it's unlikely to scale up enough to lead.

Turning now to blue hydrogen, which is frequently promoted as a decarbonisation option, we get a sense of why those doing the promoting almost always have links to the natural gas industry. While trade bodies and even respected international organisations like the IEA tend to assume carbon dioxide capture rates of 90%, real world experience suggests we would be lucky to achieve half of that. If 40% of carbon dioxide from the SMR reactor is captured, which arguably is generous based on previous results, we are still looking at emissions of more than 7.5 kg CO_2/kg H_2.

This figure is only partially due to lacklustre capture rates. The SMR route also entails emissions in the form of upstream methane leaks, which on average make up 20% of the well to gate emissions.[9] In addition, the carbon capture equipment only nabs emissions that come directly from the SMR reactor. It does not catch the carbon dioxide that is formed when methane is burned to produce heat for the process, and on average this makes up nearly 30% of the emissions from the plant.[29] In total, that's around 44% of emissions that fall outside the jurisdiction of SMR carbon capture equipment. Even if 90% of the targeted emissions were captured, that would apply to just over half of the total emissions associated with grey hydrogen—leaving a significant share untouched.

Surprisingly well named, turquoise hydrogen does sit somewhere between green and blue. If fed from fossil methane it suffers from the same upstream concerns as blue hydrogen but eliminates the process emissions and need for messy CCS. As a result, the well to gate emissions may be at a level deemed acceptable, at least as a transition technology. If fed from biomethane, the emissions can even be climate negative. The cost is also comparable to the lower end of green hydrogen's range, so any government incentives that prop up green should also work for turquoise. As stated previously, the scale of turquoise may be limited by the availability of biomethane, or quantities of solid carbon that it produces. With that caveat acknowledged, it does seem to have a place in the toolkit.

There is no perfect option. Every low-carbon hydrogen pathway involves trade-offs—between cost, climate impact, scalability, and regional feasibility. Electrolysis is clean but expensive. Blue hydrogen is cheaper but only marginally better than grey when all emissions are considered. Turquoise hydrogen may offer a middle ground, but not on a large scale. What's clear is that the current low-cost methods must be replaced—and that choosing the best alternative means looking beyond the sticker price to consider both the full emissions impact and real-world feasibility.

Big Numbers: Chapter 2

- **200 billion m³** - Natural gas used each year to produce 60% of global hydrogen
- **100 million tonnes** - Coal used each year to produce 20% of global hydrogen.
- **1 billion tonnes** - CO_2 emitted annually from hydrogen production using fossil fuels: 10 kg per kg H_2 from gas, 20 kg per kg H_2 from coal.
- **5x** - The cost of green hydrogen compared to fossil hydrogen. Clean, but expensive.
- **25%** - Emissions savings of blue hydrogen over grey, once gas leakage and real-world capture rates are factored in. Modest climate benefit at twice the price.
- **80%** - Share of carbon capture and storage projects that were cancelled or abandoned before completion. A risky, unreliable technology.

CHAPTER 3

Transporting Hydrogen

The Volume Problem

A lot of the attention given to hydrogen by politicians and the media focuses on production. Which energy sources will be used? How much will it cost? What incentives can we provide to make and use it? While these are all relevant questions and the cost of production is certainly critical, it is in transport that some of the largest impediments to a hydrogen economy can be found.

Recall that most of the hydrogen produced today is used as feedstock for chemical process plants that run continuously. In these cases, the hydrogen is produced either onsite or at a nearby location and is consumed at much the same rate that it is made. Major storage and transport infrastructure has simply not been required for the way hydrogen has been used up till now. Consequently, this infrastructure does not currently exist. If hydrogen were to become an internationally traded commodity however, then the ability to move it great distances and at low cost becomes vital.

If we wish to look at how energy is traded now, with the aim of finding a close comparison for hydrogen, the best we will do is natural gas. Like hydrogen, the energy density of natural gas under standard conditions is not high, so it must be compressed or liquified before it can be transported economically. Nevertheless, for more than half a century, natural gas has been traded over great distances and in ever increasing quantities. The scale of natural gas piping is staggering. There are currently 1.2 million kilometres of natural gas pipeline globally.[1] This is long enough to loop around the equator thirty times.

For trade routes that are not suitable for pipelines, there is an ever-growing global fleet of more than 700 vessels capable of carrying liquified natural gas (LNG).[2] Recently, both methods have been front of mind in Europe, when the Russian invasion of Ukraine forced the continent to close the natural gas pipelines from Russia and replace them with liquified shipments from elsewhere.

In 2023, 475 billion cubic meters of natural gas was traded via pipeline, along with nearly 540 billion shipped as LNG.[2] Combined, this is around 38 EJ of energy: 6% of the global energy supply. So, if natural gas can be moved around successfully, why not hydrogen? The answer lies once again in that most problematic of hydrogen's basic physical properties: low volumetric energy density. Table 3.1 compares the energy density of hydrogen with that of natural gas under some of

DOI: 10.1201/9781003361428-4

Table 3.1. Energy density comparison: hydrogen and natural gas.

	Natural Gas	Hydrogen	Ratio of Natural Gas:H$_2$
Boiling Point of Liquefied Gas (°C)	−161.6	−252.9	—
Energy Density (kWh/m³)			
Atmospheric Pressure	10	2.8	3.6
Compressed to 50 atm *Typical for moving natural gas by pipeline*	570	132	4.3
Compressed to 200 atm *Typical for moving natural gas in bottles*	2600	507	5.1
Liquid *Typical for moving natural gas by ship*	6200	2360	2.6

Note: Assumes 14.5 kWh/kg for Natural Gas.

the conditions commonly used in its transport. Natural gas is in fact a mixture of gases: predominantly methane along with several other gaseous hydrocarbons. The composition of natural gas mixtures varies by geographic location. It can even vary over time in the same location. For these reasons we cannot have definitive values for natural gas properties, but we can use representative numbers to make a good comparison with hydrogen.

We can see that at normal atmospheric conditions, a volume of hydrogen holds less than a third of the energy that an equivalent volume of natural gas contains. We also see that this is more or less true for all the conditions commonly used to transport natural gas.

If we want to substitute hydrogen for natural gas, the volume we need to send down the pipe to maintain the same flow of energy would be around four times greater. This means higher pressure in the pipe and more energy needed for pumping.

For a tank or gas bottle that you may use at home, hydrogen would only provide a fifth of the energy that the same sized bottle filled with natural gas would contain. The difference is not quite as bad in liquid form, but still not favourable to hydrogen, which has a little less than 40% of the energy that LNG holds. We would need three ship loads of hydrogen to supply the same amount of energy as one shipment of LNG, assuming all the boats are the same size.

In a nutshell, replacing a natural gas energy supply with hydrogen would require sending three to five times as much of it regardless of which method is used. This is not to say that transporting hydrogen cannot be done for technological reasons; it can and is already. However, it is more expensive to send three parcels than it is to send one and always will be. Let us now look at each transport method in more detail, for there are also interesting technical challenges that should be understood.

The Options

Road

Hydrogen can be loaded into specially designed vessels and transported by truck and trailer. This is a mature technology that has been in use for some time. However, it is viable only for relatively small quantities and distances. Hydrogen can be trucked as either a compressed gas or a liquid.

Compressed hydrogen is carried inside pressure vessels. These vessels are generally long cylinders with rounded caps. From the outside, a truck loaded with a bundle of hydrogen cylinders looks like a truck moving a stack of large drainpipes. The vessels are made of a metal or plastic inner liner to hold the gas in place, with a strong outer shell made from fibres of glass or carbon to stop the liner deforming under pressure. These fibre-composite materials have the useful combination of strength and low weight. Some common uses for similar materials include high-end bicycle frames, golf club shafts and boat hulls. Vessels made from these materials are more expensive than those made from steel, but they are also lighter and can handle higher pressure. Both properties are necessary for road transport as weight is a limiting factor for truck haulage, and higher pressures mean more hydrogen can be moved per load. A truck loaded with cylinders at a pressure of 500 atmospheres can carry about 1000 kg of compressed hydrogen.[3] However, some counties limit the allowable pressure to lower values for safety reasons.

Liquid hydrogen is transported in super-insulated cryogenic tanks. In contrast to the stacked configuration of compressed hydrogen trailers, a liquid hydrogen trailer has a single large tank that looks more like those used to transport milk or petrol. These tanks are not built to withstand high pressure, but rather to minimise heat penetration from the outside world. However, some heat will inevitably get through, causing a portion of the liquid hydrogen to boil and become hydrogen gas.

Because hydrogen gas takes up more space than liquid hydrogen, this 'boil-off' increases the pressure inside the tank, which could cause it to burst and is therefore dangerous. Boil-off could be vented into the atmosphere to release the pressure but this is undesirable both because hydrogen is a valuable product and because it is an indirect greenhouse gas. Instead, a clever work around is used: the tank is not filled completely. Leaving around 10% of the tank's capacity empty provides enough space that boil-off wont significantly increasing pressure. In this way, a large liquid hydrogen tanker can transport around 4000 kg of hydrogen; four times more than is possible with the compressed gas.

Given that so much more hydrogen can be moved in liquid form, the trucking cost of this method is significantly lower than for compressed gas. However, it costs a lot more to liquify hydrogen than it does to compress it. So, for distances less than 200 km, it ends up cheaper to move compressed gas.[4] For distances greater than this, the cost savings due to cheaper trucking justify the extra cost of liquification. In either case, transporting hydrogen by truck is only economically preferable for quantities of 10 tonnes per day or less. That is to say, the amount of hydrogen that you want to move can be handled by a small number of trucks. When larger amounts of hydrogen are being moved it is cheaper to build a pipeline.

Pipeline

Gases are transported through pipelines in much the same way you blow air through a straw—by applying pressure. Instead of lungs, pipelines use compressors to generate this pressure. A compressor is essentially a large fan, which may be powered by electricity or by burning some of the gas that is being transported. Long pipelines contain multiple compressor stations because like a child on a swing, gas eventually loses momentum and requires another push to keep moving.

Pipelines are used for both transmission and distribution. Transmission is the movement of large quantities from the production site to the region where it will be used. These sorts of pipes are large and will be made of steel. Distribution pipes are smaller and are used for transporting gas to individual users. These may be made from steel or polyethylene plastic.

As with natural gas, piping hydrogen is a mature technology. Hydrogen is transported at pressures in the range of 40–100 atmospheres, usually from a dedicated production site to a large nearby user, such as an oil refinery or chemical plant. Although most of these pipelines are only a few kilometres long,[5] in total there is currently about 2600 km of hydrogen pipeline operating in the US, 2000 km in Europe, and 600 km in Asia.[1]

It is expected that these numbers will increase substantially as hydrogen use grows. This is because pipelines are a relatively cheap and efficient way to transport large volumes of gas. Less than 2% of the energy content of hydrogen is lost per 1000 km of pipeline, which is very good relative to the other options.[3] Assuming that geography allows them, pipelines are also thought to be the cheapest transport method when the volume to be moved exceeds 10 tonnes per day.

Most of the cost in pipeline transport lies in making the pipes themselves. For this reason, there is a lot of interest in repurposing existing natural gas pipelines for hydrogen transport. After all, we need to move away from the greenhouse gas emissions associated with using natural gas, and these pipes are already built. However, the differences between hydrogen and natural gas means that this is not as simple as it might sound.

Firstly, hydrogen embrittlement is a known risk for steel pipes. Cracks can also occur in natural gas pipes due to the stresses that are placed upon them. Hydrogen embrittlement simply increases the risk, sort of how running a marathon is hard but running a marathon in full firefighting kit is that much harder. Laboratory testing has shown that even though hydrogen increases crack growth, the steel used in natural gas pipes can remain strong enough to meet the requisite engineering standards, although this is less certain for the welded joints that hold sections of pipe together.[6]

The suitability of polyethylene pipes, such as those used to carry natural gas into homes, is also uncertain. It is known that hydrogen can change the structure and properties of this sort of material but it is not yet known how this may affect the life of the pipe.[7] It has been concluded that existing steel pipes can be used if they are in good condition, although we would need to keep a closer eye on things like leaks and cracks than we normally would with natural gas.[7] Older pipes and those already showing signs of substantial cracking are probably best left out of the hydrogen business.

So, assuming a steel natural gas transmission pipeline is in good condition, can we happily send hydrogen hurtling through in its stead tomorrow? No, it's still not that easy. In addition to the pipe itself, a pipeline has many other components that play a role in moving the gas along and maintaining safe operation. Valves, flow metres, pressure regulators and the like, which have been designed for natural gas are not necessarily suitable for hydrogen.

Hydrogen may leak due to its small size, or damage the materials used in the equipment due to its different chemical behaviour. Also, because hydrogen is lighter and less energy dense, the amount of power needed to push it through is three to four times higher than for natural gas, and existing compressors may not be able to provide that.[3] If the compressors were powered by burning natural gas in turbines, these too may need adjusting or replacing because hydrogen burns somewhat differently and can produce a greater quantity of dangerous nitrous oxide emissions which would have to be accounted for with engineering design.

A recent study by the Spanish Gas Association (Sedigas) found that upgrading the natural gas network to transport a blend of 20% hydrogen and 80% natural gas would require an investment of around $750 million. Over half of this amount was for retrofitting compressor stations and replacing seals. Noting that this investment only allows a 20% blend, retrofitting clearly requires a substantial amount of money, though it's still cheaper than building a pipeline from scratch. Depending on the specific situation, it is estimated that retrofitting an existing pipeline, should it be otherwise suitable, would be somewhere in the range of 40–65% the cost of building a new one.[8]

However, given that hydrogen is an indirect greenhouse gas, leakage remains a concern. The Spanish study estimated that following the retrofit, the leakage rate would be about seven times higher than for natural gas alone. That said, we don't really have a good understanding of exactly how leaky hydrogen is. Available estimates for transmission pipeline leakage range from 0.02% to 5%.[9] If hydrogen is to be transported in large quantities and over vast distances, extra attention will need to be paid to leakage to avoid undoing any climate benefit that may be gained from its use. The difficulty in detecting hydrogen leaks is that it tends to rapidly disperse in air, rather than loitering around in amounts that can be easily measured. New technology and monitoring methods will be required to ensure the integrity of hydrogen pipeline transmission.

Ship

For distances greater than a few thousand kilometres, or routes that aren't amenable to laying pipe, shipping is thought to be the only viable method for transporting hydrogen. As always, hydrogen's low energy density by volume is a problem. The energy density of compressed hydrogen in particular is far too low and is not considered a realistic option for shipping.

Liquid hydrogen is of interest for shipping because it provides high purity product to the end user. If done well, there should be no other molecules hanging around, as would be the case with alternative shipping methods. Its main disadvantage is that liquifying hydrogen is costly and uses a lot of energy. Also, liquid hydrogen requires specially built ships to maintain its cryogenic temperature. These ships are expensive to build, when compared to conventional tankers, which adds to the overall transport cost.

Even with specially insulated tanks, some hydrogen will boil off during the voyage, or when sitting in storage tanks at either end of the journey. Good engineering can minimise boil off, or even utilise the escaping hydrogen, but cannot eliminate it completely. Lost product is therefore another downside of liquid hydrogen shipping that must be accounted for when considering options.

Liquid hydrogen is not currently shipped commercially. In fact, as of 2025, there is only one ship in the world that carries liquid hydrogen as cargo. The Suiso Frontier is a demonstration vessel that carried the first shipment of liquid hydrogen in early 2022. The hydrogen is contained in a vacuum-insulated double-walled tank, with no external refrigeration. Just like a thermos flask, this design relies on the vacuum to prevent heat flowing into the tank, keeping the liquid hydrogen at a balmy $-253\,°C$ for the duration of its journey. The Suiso Frontier's tank can hold up to 1250 m³ of liquid hydrogen, which is very small compared to commercial vessels. LNG ships typically carry between 40,000 and 200,000 m³ of product.[1]

The Suiso Frontier's name comes from the Japanese word for hydrogen. Written in kanji (水素), suiso combines the characters for 'water' and 'element.' The vessel was built specifically for the CO_2-free Hydrogen-energy Supply Chain Research Association, or HyStra for short. HyStra is a group of corporations, most of whom are Japanese, supported by the governments of Australia and Japan. The HyStra website states that its purpose is to demonstrate a hydrogen supply chain that will "lead to the dawn of a carbon-free hydrogen society".

Sadly, the demonstration value chain was very much not carbon-free. The hydrogen was made in Victoria, Australia by gasification of dirty brown coal. It was then trucked to the port where it is liquified – using electricity from Victoria's fossil-heavy grid – and loaded onto the Suiso Frontier, which runs on diesel. It's likely that the overall emissions from this supply chain are greater than simply shipping the coal from Victoria and burning it in Japan.[10]

The argument for the project is that its value is in demonstrating and perfecting the technology used in handling and shipping liquid hydrogen. Kawasaki Heavy Industries, which built the Suiso Frontier, had planned for the construction of a vessel that will carry four liquid hydrogen tanks with a combined capacity of 160,000 m³, as did their Korean competitor Samsung Heavy Industries.[1] However, in late 2024 it was widely reported that Kawasaki has scaled back their ambition and will instead focus on a 40,000 m³ ship due to the lower than expected demand for international transport of liquid hydrogen. It was also reported that they will stop taking hydrogen from Australia for development purposes and instead source it from Japan.

So, liquid hydrogen shipping remains something of a work in progress. Let's now look at how the various options compare with each other.

Comparison of Options

It doesn't take much more than a quick glance at Table 3.2 to see why pipeline is by far the preferred method for transporting pure hydrogen if all options are on the

Table 3.2. Comparison of hydrogen transport options.

Transport Method	Indicative Cost ($/MWh)	Indicative Cost ($/kg H$_2$)	Indicative Scale (t/d)	Indicative Distance (km)	Energy Penalty
Compressed H$_2$ Truck[1]	$17–23	$0.55–0.75	< 10	Up to a few 100	10–20%
Liquid H$_2$ Truck[1]	$23–45	$0.75–1.5	< 10	100s to 1000	30%
H$_2$ Pipeline (per 1000 km)[2]	$1.6–16	$0.05–0.5	> 100	10s to 1000s	2%
NG Pipeline (per 1000 km)[2]	$1.2–9	-	> 100	10s to 1000s	2.5%
Liquid H$_2$ Carrier Ship[3]	$58–75	$1.95–2.5	> 100	1000s to 10,000s	20–40%
Liquid Natural Gas Ship[3]	$10–25	-	> 100	1000s to 10,000s	15%

Sources: 1. IRENA (2022).[3] 2. DeSantis et al. (2021).[11] 3. IEA (2023).[1]

table. Trucking hydrogen starts at around six times the lower end of the price range for piped hydrogen. This is in no small part due to all the energy that is thrown away when transforming hydrogen into a denser form. Pipes lose a couple of energy percentage points per 1000 km, while trucks lose 10–30% before the key is turned in the ignition. In addition, trucks are limited by weight, meaning the quantities it can move are tiny by comparison. It's easy to see why trucks are reserved for niche applications.

Moving hydrogen by pipe is around 50% more expensive than piping natural gas, due to the differences in physical properties. Perhaps surprisingly, given our experience with natural gas transmission, its cost range is quite broad. There are a few reasons for this. Firstly, the main price contributors are the cost of steel, which fluctuates over time; and labour, which varies by location. As mentioned, substantial cost savings are also possible if existing pipelines can be repurposed rather than building new.

The cost also depends on how much gas is to be moved and consequently, the size of the pipe. Thanks to the magic of geometry, doubling the circumference of a circle results in a quadrupling of its area. This means that while doubling the size of a pipe doubles the amount of steel required, it also increases its capacity by four times. So, the more you send, the more you save. Selecting a large pipe, say 90 cm diameter, and running it at full capacity would cost around $3/MWh of hydrogen per 1000 km of length, while transmitting the same volume of natural gas in an equivalent pipe might cost a little over $2/MWh.[12]

Comparing the costs of shipping liquid hydrogen at scale is not easy because as of today, apart from the demonstration-scale Suiso Frontier, it doesn't happen. Assumptions must be made about the future efficiency and cost of the technology, based on systems that are still in development or have only been demonstrated at small scale. The indicative shipping costs presented in the table are estimates from the IEA. These costs are also the highest of the three options by a long way. Shipping liquid hydrogen is expected to be at least twice the price of shipping LNG, and using median values, it is closer to four times more expensive.

Hydrogen Smuggling on the High Seas

We have so far looked only at the options for transporting hydrogen in its pure state, but there is another way. Hydrogen can also be 'smuggled' in some other substance. This involves converting the hydrogen into said carrier substance for transport and reconverting back into hydrogen at the destination. Many different carriers have been investigated and it continues to be an active area of research. The primary challenge is that these conversions inevitably have a price, both in terms of money and energy. As a result, they make little sense for the small volumes associated with trucking and are less convenient than simply using the real thing when it comes to pipes. However, smuggling is of interest for shipping, where distances are long, and volume is limited.

The first option is to react hydrogen with other atoms to form a different chemical that is easier to handle and can be deconstructed back into hydrogen at the point of use. Potential chemical species for this method include methanol (CH_3OH), and ammonia (NH_3). Both are internationally traded chemicals now, so we have some infrastructure and experience handling them, albeit carefully, as they are both highly toxic. Ammonia and methanol production are among the largest current users of hydrogen, and replacing the grey hydrogen they rely on is itself a potential use for green hydrogen. This will be described further in Chapter 5. For now, it is enough to say that methanol requires a fossil-free source of carbon dioxide, and ammonia doesn't. Consequently, clean ammonia is expected to be cheaper to produce and easier to scale than clean methanol. This gives it the edge as a hydrogen smuggler.

As an alternative to forming an entirely new molecule, hydrogen can also be made to attach itself to carefully selected materials. Attachment can be either by chemical reactions that form strong bonds between hydrogen and the material, or by weaker chemical interactions where the hydrogen 'sticks' to the material's surface instead. In either case, the idea is that hydrogen gas can be loaded and unloaded from the carrier material by changing some combination of its temperature and pressure. Conceptually, this is a bit like loading a sponge by placing it in water and then unloading it by squeezing.

For decades, metals have been investigated as potential hydrogen carriers. Certain metals can chemically bond with hydrogen, forming compounds that are known, logically enough, as metal-hydrides. It is now understood that light-weight metallic elements like sodium, lithium, magnesium and aluminium tend to work best.[12]

A key difficulty with metal hydrides is creating enough parking spaces for the hydrogen. You have probably seen piglets at feeding time, desperately scrambling over each other to secure one of their mother's teats. Once all the spots are full, the piglets that got in early have effectively formed a wall, blocking out the others. The situation is similar with metal hydrides; the first hydrogen atoms to react with the metal form a metal-hydride barrier that prevents other hydrogen molecules from getting through and limits the amount that the metal can hold. For example, magnesium hydride

(MgF_2) theoretically contains 7.6% of hydrogen by weight, but in practice it is much less because of this barrier effect. Advanced materials engineering techniques can be used to create nano-scaled structures that contain more 'teats', but these techniques are expensive and don't easily translate to a commercial scale. A further problem for metal hydrides is that the loading and unloading cycle is energy intensive because both steps require high temperatures. As a result, the amount of energy available as hydrogen tends to be a fraction of what was used to get the hydrogen to the end user.

An alternative class of materials known as liquid organic hydrogen carriers (LOHC) are believed by some to hold more promise. There is currently no consensus on which specific substance is the best LOHC, and there have been many proposals. LOHC are generally made up of hydrocarbon molecules that form as rings. These are known as 'aromatic' compounds and have names that are a bit of a mouthful. Two of the more commonly invoked LOHC are naphthalene and methylcyclohexane.

Aromatic rings are useful because the carbon atoms within them can have one or two chemical bonds. In other words, they have an extra pair of hands that they can use to grab onto hydrogen molecules. They can also release the hydrogen again when coaxed and thus are able to be loaded and unloaded. LOHC are mostly derivatives of oil, which is both useful and a problem. It is useful because like other oil products, they are liquid at ambient temperature and so can easily be handled using our existing fossil fuel infrastructure. It is a problem because the whole reason we are looking at alternative energy is the need to move away from oil products and there is currently no way to make them at scale without oil.

LOHC are also very expensive to make. At the lower end of the cost range, the price of a shipload of LOHC is almost as much as the cost of the ship that carries it.[3] In addition, some LOHC is lost to the environment during the loading/unloading cycle and would need to be replaced. LOHCs also share the challenge of metal hydrides in that they don't really hold all that much hydrogen: in the range of 4–7% by weight.[3] So, a container ship would be traipsing back and forth between the hydrogen producer and user carrying a big tank of oil, with a little bit of hydrogen attached. That's not going to earn many stars for energy efficiency. Nevertheless, LOHC are often short-listed alongside liquid hydrogen and ammonia as the least-bad means for shipping hydrogen.

Let's now examine Table 3.3, which is again populated by projections from the IEA. These numbers are estimates of the 2030 cost for the three methods. The assumed shipping distance is 8000 km, which is roughly the distance between Brazil and northern Europe, or the east coast of Australia and Japan/Korea. These numbers include the cost of building and running the ships, storage tanks, and process equipment. They also assume improvements in efficiency compared to the present time. For example, the energy loss due to liquifying hydrogen is 20% in the table, rather than the 30–36% that it is today. This is due to assumed technological improvements and increasing the size of liquefaction plants, so there is a little bit of optimism priced in. For comparison, liquifying natural gas typically uses around 10% of its energy content.

Recall that LNG costs somewhere between $10–25 per MWh to ship. Even with optimistic assumptions about the future, we can see that shipping hydrogen is still

Table 3.3. Monetary and energy cost of shipping selected hydrogen carriers.

Shipping Cost	Liquid Hydrogen		LOHC		Ammonia	
	Monetary ($/kg H_2)	Energy Loss	Monetary ($/kg H_2)	Energy Loss	Monetary ($/kg H_2)	Energy Loss
Making hydrogen carrier	$0.80	20%	$0.25	5%	$0.60	14%
Moving hydrogen carrier	$1.70	≤ 7%	$0.75	≤ 3%	$0.40	≤ 1%
Reconverting to hydrogen		≤ 1%	$1.05	36%	$0.95	22%
Total	$2.50	20–28%	$2.05	41–44%	$1.95	36–37%

Source: IEA (2023).[1]

very expensive by comparison. Liquid hydrogen is the most expensive method at around $2.5/kg H_2 or $75/MWh of energy. The persistent costliness is due partially to the cost of liquefaction, and the energy that is lost in the process. However, the biggest expense is expected to be the cost of building the ships with their giant thermos-like vacuum storage tanks. IRENA estimates that once the technology is mature, sometime around 2035–2040, a liquid hydrogen tanker will cost five to seven times as much as a vessel designed for carrying ammonia or LOHC.[13] In dollar terms, that is up to half a billion dollars for a liquid hydrogen tanker compared to $50–100 million for an ammonia tanker that will carry about the same amount of energy.

Neither ammonia nor LOHC are as costly as liquid hydrogen across the first two steps: making and moving the hydrogen carrier. In fact, they both cost about $1/kg H_2 for these steps combined. Where the big expense comes is reconversion back into hydrogen. Cracking ammonia back and unloading LOHC both require high temperatures, and the energy required to produce these temperatures constitutes a good amount of the energy contained in the hydrogen. Furthermore, this heat is demanded by the laws of physics and is not something that can be waived away with better technology. This requirement equates to increased monetary costs because when the product is energy, any conversion loss means that some product is literally being discarded on route to the customer. For both ammonia and LOHC, the cost of hydrogen reconversion is about half of the total shipping cost. Of the two, ammonia is the most cost effective overall but even $1.95/kg H_2 ($59/MWh) may be a clean energy deal breaker.

It is important to note that if cracking isn't required, that is to say, if ammonia can be used to fulfil whatever role was intended for the hydrogen, then the shipping cost becomes much more affordable. Removing the cracking steps results in a shipping cost for ammonia of around $30/MWh for an 8000 km voyage. This is still more than the $10–25 per MWh that we pay for shipping LNG today, but for customers that are currently paying the higher end of the range, a $5/MWh green premium may be acceptable. The energy consulting firm Wood Mackenzie has reported that more than 85% of developments aiming to make and export low-carbon hydrogen, plan to ship it as ammonia, suggesting that this is indeed the case.

So, if transporting hydrogen at scale is untenable by ship, what does this mean for dreams of an international trade in green energy? Could ammonia be the green molecule we are looking for?

Big Numbers: Chapter 3

- **3 - 5x** - Volume of hydrogen needed to match the energy of natural gas. A bulky energy carrier, hydrogen means making, moving, and storing a lot more gas.

- **33%** - Energy lost when liquifying hydrogen due to –253°C boiling point. An expensive and wasteful prerequisite for shipping.

- **0** - Number of commercial liquid-hydrogen carrier ships that have been built. An expensive technology with uncertain demand.

- **85%** - Share of export-oriented green hydrogen projects planning to ship as ammonia, to sidestep hydrogen's transport challenges.

- **2%** - Energy lost per 1000 km of hydrogen pipeline transmission – far more efficient and straightforward than trucking (10–30%) and shipping (20–40%).

CHAPTER 4

Hydrogen Derivatives

◇◇

The Ammonia Alternative

It's Mr Drumm's year 12 chemistry class. We've just been presented with a large glass bottle containing a clear, colourless liquid. We are invited to smell the liquid and instructed to do so by waving a hand over the opening to catch just a whiff of the vapour. One lad – let's call him Bushy – was not paying sufficient attention to the instructions and placing his nose directly over the bottle, took a generous snort.

Instantly, his head flung back. Bug eyed and face turning red, he shot out of the classroom in pursuit of fresh air, followed closely by the laughter of his classmates. Bushy was fine, and we all received an indelible lesson on the pungent potency of ammonia.

Constructed from one nitrogen atom and three hydrogen atoms, ammonia (NH_3) is a highly reactive, colourless gas that is quite poisonous at high concentration. Often referred to as ammonia solution, the liquid in the bottle was actually ammonium hydroxide (NH_4OH): an alkaline solution of ammonium (NH_4^+) and hydroxide (OH^-) ions, formed by ammonia gas dissolving in water.

The nature of the ammonia molecule is to do just that. It is highly hydrophilic, which means it eagerly seeks out water. When it finds some, ammonia will steal a hydrogen ion like so:

$$H_2O + NH_3 \rightarrow NH_4^+ + OH^-$$

water + ammonia → ammonium ion + hydroxide ion

This tendency is both a great boon, and a great danger. Ammonia is not fussy about where it finds water, and because our bodies are 55 to 60% H_2O, it's best not to share a room with the stuff. At a concentration of 100 parts per million (ppm) ammonia can cause irritation to our eyes and throat. For comparison, carbon dioxide is around 400 ppm in air (and rising), so even four times less ammonia can be a nuisance. At concentrations in the 500–700 ppm range, ammonia can cause chemical burns to the throat, lungs, eyes and skin. Normally our skin is a pretty good barrier to water-based substances but ammonia reacts with the fats and oils in skin, essentially turning them into a sticky soap that does little to block further ingress of the gas.[1] The ability to break down fats and grease is actually why ammonium is found in many household

DOI: 10.1201/9781003361428-5

cleaning products and why wearing gloves when using them is recommended. At even higher concentrations, ammonia can cause death by suffocation as the chemical burns and swelling shuts down the respiratory system.

It is therefore ironic that this nasty substance is also vital to life on Earth. Ammonia is an integral part of the nitrogen cycle. It is produced naturally in soil when microbes chow down on rotting organic material. When water is about – as is often the case in soil – ammonia will transform to the ammonium ion, which is the vehicle by which plants receive nitrogen. Nitrogen is an essential building block of organic structures. It is a component of DNA as well as amino acids which are a structural component of cells. Put simply, without sufficient nitrogen, plants can't grow. So, for the existence of plant life, we can be thankful for ammonia's unquenchable thirst. This is the great boon.

During the 20th Century it was found that direct application of ammonia was more effective at promoting plant growth than feeding dung to the soil microbes. In fact, synthetic fertilisers were one of the pillars of the so-called Green Revolution that saw major leaps in the productivity of agriculture starting in the 1960s. It is now estimated that around half of all humans alive on earth today are reliant on synthetic fertilisers, many of which are derived from ammonia.

The means of industrial ammonia production has not changed much since its invention in the early 20th century. The Haber-Bosch process, which will be described further in Chapter 5, utilises high temperature and pressure to combine hydrogen with nitrogen captured from the air. Between 70 and 80% of the roughly 180 million tonnes of ammonia produced each year is turned into nitrogen fertilisers. These are used to grow food crops for both humans, and the domesticated animals we like to eat. Without this fertiliser, the global population could not have reached the size it is today. The remaining ammonia is used to make explosives, or as a feedstock for other chemicals.

Approximately 10% of ammonia production is traded currently.[2] The US has around 3000 km of ammonia pipeline, and there was a 2400 km pipe between Russia and Ukraine, but this was damaged in 2022 during Valdimir Putin's "special military operation" according to reporting from Reuters. More commonly, ammonia is transported by ship. There are currently over 120 ports with ammonia terminals and a similar number of ships capable of carrying ammonia.[2] The ammonia industry is mature, which means we have a lot of experience making it and moving it about. This industry is also built entirely upon the value of the nitrogen atom. In a low-carbon world, might ammonia also be prized for the value of its hydrogen atoms?

In the previous chapter we noted that shipping liquid ammonia made from hydrogen is a more cost-effective way of moving energy then shipping liquid hydrogen, as is evidenced by 85% of clean hydrogen export projects planning to use the ammonia option. Shipping as ammonia was estimated to cost around $30/MWh and 15% of the energy contained in the hydrogen that was used to make it: much less than liquid hydrogen. Let us now look at why this is. A comparison of relevant physical properties of ammonia and hydrogen are provided in Table 4.1.

Even though hydrogen has nearly six and a half times the energy content of ammonia on a weight basis, its low density once again makes it the poor cousin

Table 4.1. Physical properties of liquid hydrogen and ammonia.

	Specific Energy (kWh/kg)	Density (kg/m³)	Energy Density (kWh/m³)	Boiling Point (°C)	Ship boil-off (%/day)
Liquid Hydrogen	33.3	70.9	2360	−252.9	0.3[1]
Liquid Ammonia	5.2	682	3550	−33.3	0.025–0.05[2]

Sources: 1. Kawasaki Heavy Industries Ltd (2023). 2. Al-Beiker (2023).[3]

when storage space is a limiting factor. Ammonia molecules can pack in much more tightly than hydrogen molecules, so a storage tank can hold nearly ten times more of them. Curiously, that tank of ammonia would contain 70% more hydrogen than an equivalent tank of liquid hydrogen. The ammonia molecule is only 17.76% hydrogen by weight, but the difference in density more than makes up for this, resulting in more hydrogen overall. Consequently, the energy density of liquid ammonia is greater than liquid hydrogen by about 50%.

In addition to providing more energy per shipload, ammonia is also a better traveller. This is because keeping the passengers comfortable is far easier when they require a −33°C cabin, than when they demand −253°C. This massive difference in boiling pt manifests in the relative boil-off rates. Remember that boil-off is the gas which is formed from the liquid product when heat infiltrates the storage tank. The Suiso Frontier, currently the only liquid hydrogen tanker in existence, has a boil-off rate of 0.3% per day, according to its builder, Kawasaki Heavy Industries. Ammonia's rate is a fraction of this, but for comparison, let's use 0.05% which is at the higher end of the range.

To illustrate the effect of boil-off, let's assume we have two ships, each with a capacity of 100,000 m³, and we fill one with liquid ammonia and the other with liquid hydrogen. The hydrogen ship will hold 7090 tonnes, which is equivalent to 226 GWh of energy. This amount of energy is approximately the Tokyo region's electricity consumption over seven hours. The ammonia ship will hold 68,200 tonnes, or 352 GWh: equivalent to 11 hours of Tokyo's bright lights. Now if we assume a two-week long voyage, say from Australia to Japan, the boil off losses for hydrogen would be around 292 tonnes (9.7 GWh), while the losses for ammonia would be 476 tonnes (2.5 GWh). So not only is the hydrogen ship setting out with less energy, but it's also losing more on the way. In this scenario the customer would receive 350 GWh of ammonia and 216 GWh of hydrogen; 62% more energy is delivered using ammonia.

It is important to stress that much of this efficiency advantage is lost if the ammonia is cracked back into hydrogen. Instead of delivering 62% more energy following a hypothetical 8000 km voyage, cracked ammonia would only provide around 25% more energy than liquid hydrogen. To take full advantage of ammonia's greater energy density, we need to be able to use it as ammonia rather than hydrogen. The question then becomes, can ammonia be used as a fuel?

Theoretically, ammonia can be used to generate electricity, either by burning it in a thermal power plant or by feeding it through a fuel cell. It can also be burned in an internal combustion engine to produce mechanical power. In this way, its versatility as a fuel is similar to hydrogen and natural gas. However, there are challenges to using ammonia as a fuel in addition to its hazardous nature. Ammonia requires a relatively large amount of energy to ignite, which is good for safety but problematic when used as a fuel. It also has a low flame velocity, which is the speed that a flame moves through a mixture of fuel and oxygen and relates to the rate of heat production. At best, the flame velocity of ammonia at standard temperature and pressure is 7 cm/s, which is five times slower than natural gas at 35 cm/s.[4] Interestingly, hydrogen's flame velocity (300 cm/s) is at the other end of the spectrum. Its ability to rapidly release energy is why it is used in rocketry.

Low flame velocity is a problem in combustion equipment because, like trying to light a campfire on a windy day, it is difficult to maintain a stable flame. Flame stability is essential for the safe and efficient operation of engines and burners. Basically, ammonia is a slow swimmer and cannot be used in a combustion system designed to run on conventional fuels.

The solution to this problem is usually to mix ammonia with a better burning substance like hydrogen, methane, or diesel. This enabler is termed a 'pilot fuel'. Obviously, the need to handle two fuels rather than one is not ideal in many applications. This is especially true when the desire to avoid problems associated with the pilot fuel was the whole reason for using ammonia to begin with. There are other engineering tricks that can be applied to improve the flame characteristics of ammonia such as using burners that create swirl patterns or increasing the temperature of the combustion chamber. Unfortunately, methods that improve the flame stability also tend to exacerbate the second major challenge of ammonia combustion: high nitrogen oxide emissions.

While it's true that there will be no carbon dioxide emissions from burning ammonia (unless the pilot fuel is a hydrocarbon) it is also true that there will certainly be a (un)healthy amount of nitrogen oxide. Recall that NOx is a major contributor to air pollution and a cause of respiratory problems, and form whenever combustion happens in air at sufficiently high temperature.

When air is the source of the nitrogen, like in a car engine, these emissions are called 'thermal NOx'. Ammonia however is predominantly made up of nitrogen. Burning ammonia therefore not only risks thermal NOx, but it also *guarantees* 'fuel NOx' where the source of nitrogen is the fuel molecule itself.

Ironically, the cure for NOx is typically more ammonia. Ammonia reacts with NOx, through several different chemical reactions, producing nitrogen gas and water. If excess ammonia is fed to a combustion chamber – meaning, there is more ammonia in the chamber than there is oxygen – much of the NOx will be eliminated. Unfortunately, this means emitting a lot of unburnt ammonia as well, which is clearly not what we want because it's dangerous and wasteful. Alternatively, if the process is large enough to justify the expense, a selective reduction unit (SCR) can be installed. Essentially, this is an end of pipe cleaning unit that catches the exhaust gases and reacts them with ammonia over a catalyst. The downside of this approach is that the ammonia used to clean up the mess cannot also be used as fuel, so the efficiency of the operation is reduced.

However, it's fair to say that it is still early days when it comes to our understanding of ammonia combustion. While there has been a lot of laboratory scale work done, pilot-scale investigations are still recent and there are no industrial-scale systems commercially available at this time.[4] The same can be said for ammonia fuel cells. Hydrogen fuel cells are now commercially available in a few 'flavours' such as alkaline, proton exchange membrane, and solid oxide. In contrast, fuel cells that can be fed directly with ammonia are still in the very early stages of development.[5] Compared to hydrogen, the ammonia reactions in a fuel cell are slower and rather more complex at moderate temperatures.[6] Further research is needed to find the combination of electrode materials and electrolyte that can meet the key requirements for commercial ammonia fuel cells: high efficiency, safety, and long operational life.

Our ability to convert ammonia into electricity, or rather our current lack of it, is probably the biggest reason why the idea of an ammonia economy is not as popular as that of the hydrogen economy. If this is remedied in the future, then use cases requiring storage and transport of renewable energy may well open to ammonia.

However, a second very good reason why ammonia's uses are limited is its very toxic nature. Hydrogen is explosive, which is not to be taken lightly, but it won't burn our lungs if we are accidentally exposed. Ammonia handling will always require carefully engineered systems and well-trained operating personnel. It is far from an ideal energy carrier, but in some situations, it is seen as a less bad option than hydrogen. Ammonia is currently receiving the most attention for its use as a shipping fuel, and as a blending agent in coal-fired power stations. These use-cases will be described in the following chapter.

Electrofuel

In contrast to ammonia, which is pitched as a new kind of energy carrier, synthetic fuels use hydrogen to rebuild the same old fuels we've been using for decades.

Electrofuel, which also goes by the handle 'e-fuel', is made using electrolytic hydrogen. With a skyline of shiny columns and piping, e-fuel plants resemble mini oil refineries, and their products are very close to traditional fossil fuels at the molecular level. In fact, they can be tailored to suit any mode of transport. There's e-gasolene and e-diesel for the automotive sector, e-kerosene for aircraft, and e-methanol – a potential shipping fuel. These products can often be 'dropped in' to existing engines and infrastructure without the need for modification: a major advantage for the automotive and aviation sectors in particular.

It is important to note here that an e-fuel is not necessarily a green fuel. The use of green hydrogen in its production is necessary but not sufficient to make a synthetic fuel low-carbon. This is because using an e-fuel in a combustion engine releases greenhouse emissions just as a fossil fuel would. At best, an e-fuel can be carbon-neutral—but only if the carbon fed into the process was first captured from the atmosphere. This way, the carbon dioxide that is released into the atmosphere when the fuel is burned is simply returning to where it came from. In contrast, burning a fossil fuel releases carbon that was previously stored underground, increasing the concentration of CO_2 in the atmosphere.

Carbon is most commonly fed into an e-fuel process in the form of carbon dioxide. This could be sourced from the atmosphere by direct air capture machines, but this is very expensive; currently \$600/t CO_2 or more.[7] More realistically, carbon dioxide will be captured from the exhaust stream of an industrial process. If the process' raw material was biogenic – meaning it captured atmospheric CO_2 while it was growing – the exhaust stream may be appropriate for carbon-neutral fuel. One example would be a brewery, or ethanol plant, both of which use fermentation of crops to produce alcohol. However, a process that generates carbon emissions from a fossil stock – like a power plant burning coal – cannot be considered suitable for carbon neutral fuel, even if there is no technical reason why the CO_2 could not be used.

Keeping in mind the requirements for carbon neutrality, lets now look at how e-fuels are made. As mentioned, e-methanol is a potential shipping fuel, but its production will be described in the following chapter, because it is relevant to decarbonization of the wider methanol industry. We will instead focus here on the making of conventional hydrocarbons, like jet-fuel and gasolene, which happens a little differently.

These fuels use the Fischer-Tropsch (FT) process, which was developed in Germany following World War 1. During this time, the recently defeated Germans were obliged to pay reparations to the Allies and faced restrictions on access to natural resources. Both interventions contributed to a shortage of liquid fuels in Germany.

Working together at the Kaiser Wilhelm Institute for Coal Research, the applied chemist Franz Fischer and catalyst expert Hans Tropsch successfully invented a process for synthesising liquid fuels from Germany's abundant stocks of coal. Employed initially to help circumvent the economic constraints, Germany's use of FT was strategically scaled up during the 1930s under the Nazi regime. By the 1940s, the technology was crucial in enabling Germany to sustain its military operations despite Allied blockades that limited access to oil.

After WWII, interest in the Fischer-Tropsch process waned as access to crude oil increased and the process remained expensive compared to conventional oil refining. However, it found new life in the 1950s in South Africa. With apartheid-era sanctions limiting oil imports, the state-owned Sasol (South African Coal, Oil, and Gas Corporation), adapted and improved FT technology to create synthetic fuels from its plentiful coal reserves. While it has in the past been something of a pariah-state process, FT is now being revisited in the search for alternative, low-carbon fuels.

The Fischer-Tropsch process calls for carbon monoxide and hydrogen as its main ingredients. However, carbon monoxide is not readily available at industrial scale, so it's produced using hydrogen to reduce captured carbon dioxide. This reaction requires heat energy to proceed, so the reactor runs at a temperature of 600–800°C in the presence of a catalyst. The overall reaction is as follows:

$$CO_2 + H_2 \rightarrow CO + H_2O$$

carbon dioxide + hydrogen → carbon monoxide + steam

Once dried, the carbon monoxide is then combined with a fresh hydrogen feed in a Fischer-Tropsch reactor, along with another catalyst. Through a series of reactions, the hydrogen and carbon monoxide link up to form long-chain hydrocarbons. The length of the chain is influenced by the type of catalyst used, the temperature and pressure of the reactor, and the ratio of the ingredients.

Liquid fuels are made up of hydrocarbon molecules of varying sizes. Gasolene is the lightest because it has the smallest molecules: 5 to 12 carbon atoms each. Kerosene is a little heavier with 10 to 15 carbon atoms, while diesel is the heaviest of the three with 10 to 20. The example Fischer-Tropsch reaction below uses decane – the smallest chain typically present in kerosene:

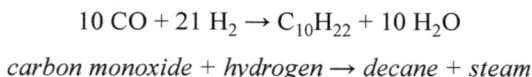

$$10\ CO + 21\ H_2 \rightarrow C_{10}H_{22} + 10\ H_2O$$

carbon monoxide + hydrogen → decane + steam

The product of the FT reactor must again be dried and then further refined to obtain the desired products. Refining is required because the FT reactions can be influenced but not controlled completely. For example, a process optimised for jet fuel will still produce a large quantity of gasolene. Altogether, the Fischer-Tropsch process requires approximately 50 to 55 kilograms of hydrogen per tonne of e-fuel produced.

As you may sense from this process outline, the main drawback of e-fuel is its high cost of production. Table 4.2 breaks down the cost of e-kerosene made under ideal conditions: a large-scale plant, access to high quality renewable electricity, and high-concentration biogenic carbon dioxide. Other e-fuels, such as e-gasoline and e-diesel, are expected to have similar costs, since the underlying process is the same for all.

Even with the best-case assumptions, the IEA estimates the cost of e-kerosene today at around $80/GJ, which is 4 to 5 times higher than the price of conventional jet fuel. Under less ideal conditions, for example a lower concentration carbon dioxide source, the e-kerosene price could be 50% higher still.

If we assume that in the future the cost of electrolysers drops by 60% and the cost of electricity decreases by a third, neither of which is assured, e-kerosene would still be 2–3 times more expensive than fossil-kerosene. Furthermore, Table 4.2 shows that the cost of capturing carbon dioxide is only a few percent of the total e-kerosene

Table 4.2. Production cost estimates for E-kerosene.

Component Cost	Ideal Plant Today ($/GJ)	Optimistic 2030 Plant ($/GJ)
Electricity	30	20
Electrolysers	30	10
FT Synthesis	16	16
Water	1	1
CO_2 Capture	3	3
Total Cost	80	50
Jet Fuel Cost Today ($/GJ)	17–23	

Source: IEA (2021).[8]

cost. Making the green hydrogen and the Fischer-Tropsch process itself are the big expenses.

Cost, however, is not the only obstacle. Scaling production depends on access to biogenic CO_2, which remains severely limited. Current capture rates are around 2.5 million tonnes per year, primarily from ethanol plants in the United States. According to the IEA, replacing just 10% of aviation fuel with e-kerosene would require 200 million tonnes per year—a target well beyond foreseeable supply. On current trends, biogenic CO_2 availability is expected to reach only 90 million tonnes by 2030.[8] Even the IEA's optimistic *Net Zero Emissions Scenario* projects just 160 million tonnes, which could be considered a best-case outcome. High costs may clip the wings of e-fuels; lack of feedstock could cut them off at the knees.

Despite the gruesome economics, early pilot projects are in the works. The first integrated e-fuels plant, according to its developer, began operation in 2022. Located in southern Chile, the Haru Oni plant uses a single wind turbine to produce green hydrogen, which it combines with CO_2 trucked in from a brewery to make up to 130,000 litres of e-fuel per year, according to its developer. It had originally planned to supply CO_2 using direct air capture, but industry reporting suggests that this equipment was only brought in at the end of 2024.

Its first shipment of e-gasolene was made in late 2022 to Porsche in Europe. Porsche is also an investor in the project and is reported to have used the fuel in its experience centre driving ranges. What it has not said is how much the fuel cost, but the Potsdam Institute for Climate Impact Research estimated the price at $50 per litre, which is a hundred times higher than the typical wholesale price of gasolene.[9] Given that Porsche was the buyer, the jokes write themselves.

To be fair, the Potsdam institute also wrote that increasing scale could see the price drop to $2 per litre ($58/GJ) if the plant followed through on planned expansions. However, this is still four times the typical gasoline price so it is doubtful whether that will ever happen, and whether Haru Oni becomes anything more than a small-scale proof-of-concept operation.

Closer to the demand centres of Europe, and running a little behind schedule, Germany's largest e-fuel pilot plant is nearing completion. Ineratec's pioneer power to liquid fuel plant, will be based in a Frankfurt industrial park from which it will receive biogenic CO_2. According to its developer, it is expected to produce up to 2500 tonnes of e-kerosene per year for use in aviation. To put this in perspective, a return flight from London to New York in a 747/A380 size airliner uses something like 150 to 180 tonnes of fuel. So, the pilot plant's annual production could fuel around 15 return flights, which is less than a day's worth of demand.

At a somewhat larger scale, construction has begun in China on what the local government has called the world's first 100,000 tonne green aviation fuel demonstration project. Located near the city of Shuangyashan in eastern China, the plant is said to combine CO_2 from biomass gasification with hydrogen produced from nearby wind and solar, to make 100,000 tonnes per year of e-SAF by 2027. Subsequent expansion is planned to increase the capacity to 300,000 tonnes per year of e-SAF and 200,000 tonnes of green methanol.

However, intentions are not always realised, even after construction has begun. The FlagshipOne project, situated in Sweden, was to be Europe's largest e-methanol plant. The developer Orsted had intended to supply 50,000 tonnes of e-methanol per year to the shipping industry for fuel. Construction began in May of 2023, but was stopped fifteen months later when the project was cancelled, reportedly due to a lack of buyers. Without guaranteed buyers for the fuel, the Orsted felt it could no longer justify investment in the plant. This underscores the challenge of bringing costly e-fuels to market and highlights why we should take a 'believe it when I see it' position on claims about future production.

HEFA

Even if e-fuels remain marginalised due to their high cost, hydrogen will most likely still be in demand for the production of certain types of biofuels. Biofuel is a wide umbrella term that refers to any fuel that is made from biomass: organic material such as plants, algae, and animal waste. While perhaps not strictly a hydrogen derivative, the class of biofuel that is known by the acronym HEFA (hydrogenated esters and fatty acids) does require hydrogen as a key ingredient: around 20–45 kilograms of hydrogen per tonne of fuel.[10] For this reason, we include it in the discussion here alongside e-fuel.

HEFA is currently a minority biofuel; the two most prominent being ethanol and biodiesel. Neither of these require hydrogen in their production, although small quantities may be used to improve their quality. They are described here simply to provide context.

Ethanol is by far the most voluminous biofuel, with global production exceeding 100 billion litres annually. It is made by fermenting sugar derived from crops. The United States and Brazil are the largest producers, with the US using corn as its primary feedstock and Brazil relying on sugarcane. This has been a source of controversy, with many arguing against the use of crops that could be used for food. Though sometimes referred to as bioethanol, fermentation is in fact the dominant commercial production method, making this term somewhat redundant. Ethanol is typically blended with conventional transport fuels to reduce their carbon intensity, although in Brazil it is used directly as a fuel in modified engines.

Biodiesel is the second most voluminous biofuel with production around 45 billion litres per year. It is made by reacting alcohol (methanol or ethanol) with vegetable oils, animal fats, or used cooking oil. It is usually blended with fossil diesel to reduce overall emissions and because it is not quite suitable to be used as a drop-in fuel. This is because its physical properties can lead to clogging or blockages in unmodified engines under certain conditions.

The main products of HEFA are the rather confusingly named 'renewable diesel' and 'biojet fuel'. The term 'biodiesel' is reserved for the products of the alcohol reaction route mentioned in the previous paragraph. HEFA uses the same oils and fats as biodiesel, but instead of reacting them with alcohol, it uses hydrogen at elevated pressure and temperature. The role of the hydrogen is to displace oxygen atoms from the fatty molecules, transforming them into long-chain hydrocarbons. A simplified hydrogenation reaction is provided below:

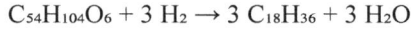

$$C_{54}H_{104}O_6 + 3\ H_2 \rightarrow 3\ C_{18}H_{36} + 3\ H_2O$$

triglyceride (fat/oil) + hydrogen → octadecane (diesel/jet fuel) + water

Unlike biodiesel, products of hydrogenation could be drop-in fuels for some applications. Renewable diesel is a higher quality product than biodiesel, offering superior performance and stability. However, biodiesel is cheaper and is currently made in much larger quantities. Renewable diesel is produced at a rate of 13 billion litres per year: a third of biodiesel's volume.

Although they use the same process, renewable diesel is currently produced in much higher volumes than biojet fuel. Only 200 million litres of biojet were made in 2022 – less than 2% of renewable diesel's production.[11] This is largely because hydrogenation chemistry favours the formation of molecules that fit within the diesel fraction. Only around 15% of the reaction products fall within the jet fuel range.[12] Additionally, automotive fuels were an early focus of government incentive schemes, which often subsidised the cost of diesel production.

It is possible to optimise the hydrogenation process to increase the proportion of biojet to near 75%, albeit with higher production costs and perhaps lower overall output.[8] Government interest in reducing the carbon intensity of aircraft has also increased in recent years, at least in the rich nations of the EU, the US, and Japan. This interest has resulted in blending mandates, which will come into force soon and incentivise more biojet production.

As with other low carbon fuels, HEFA are a lot more expensive to make than their fossil analogues, with the production cost heavily influenced by the cost of the feedstock. Table 4.3 summarises the production costs range along with carbon footprint and abatement cost for several aviation fuel options. As with e-kerosene,

Table 4.3. Production cost, life cycle emissions, and carbon abatement cost of aviation fuels.

	Production Cost ($/GJ)	Life Cycle Emissions (kg CO_2/GJ)	Carbon Abatement ($/t CO_2)
Fossil Jet Fuel	14–16[1]	90 [2]	-
HEFA Fuels today	27–39[2]	12–19 (waste oils) [3] *79–89% reduction*	140–350 (waste)
			160–450 (crops)
HEFA Fuels 2030	24–36[2]	23–34 (crop oils) *62–74% reduction*	100–310 (waste)
			120–390 (crop)
E-kerosene today[2]	80	2–25 *72–98% reduction*	700–1000
E-kerosene 2030 [2]	50		400–550

Sources: 1. IEA (2021).[13] 2. IEA (2024).[8] 3. Xu et al. (2022).[14]

which is also included for comparison, there is little difference in costs of renewable diesel and biojet as the underlying process is largely the same. We can see that fossil jet fuel remains by far the cheapest option today. Biojet is two to three times higher, which appears a bargain next to e-kerosene. In both cases, the situation would be improved if the hopes for cost reductions are achieved by 2030, but only by a little. These fuels will remain horrendously expensive for the foreseeable future.

In terms of straight emissions reduction, we can see that e-kerosene has the highest ceiling with near zero emissions as a best-case outcome. Not too far behind is HEFA produced from waste oils such as animal tallow or used cooking oil. Providing carbon reductions up to 89%, this route is a step up from HEFA that uses oil from energy crops. This is largely because cultivating crops such as soy and canola requires fertilizer, land, and energy for farming equipment, all of which add upstream emissions. Waste oils on the other hand, are considered as byproducts with minimal additional environmental burden.

Combining costs and carbon intensities lets us estimate the 'carbon abatement cost' of hydrogen-based fuels—the price tag for cleaning up a tonne of CO_2. When that price is high, it's a sign the solution might not be worth the money. As we saw in Chapter 2, firms today typically pay less than $50 per tonne for the right to pollute, and only in a few rare cases is the cost above $100. If paying the penalty is cheaper than switching fuels, that's almost certainly what they'll do.

From Table 4.3, we can see that at best, HEFA fuels can deliver an abatement cost in the ballpark of $100–150 per tonne—potentially an economically rational choice in regions with the highest carbon prices, especially if those prices rise. But at the upper end of the HEFA cost range, the case becomes much shakier. Looking even more hopeless is e-kerosene, which even under optimistic future assumptions carries a carbon abatement cost of at least $400 per tonne. At those levels, no firm is going to switch voluntarily. If e-fuels are to have a business case, governments will need to mandate their use.

The basic problem is clear: hydrogen-derived fuels can more or less match fossil fuels in function, but not in cost. The chemistry is not the barrier—the economics is. While small volumes of HEFA and e-kerosene may find buyers where emissions reductions are heavily subsidised or mandated, these fuels will not drive a market-led energy transition. In practical terms, hydrogen-based fuels look set to play niche roles in sectors like aviation and shipping, where alternatives are few, rather than become mainstream substitutes for petroleum products.

Yet hydrogen's role in transport is only part of the story. In the following chapters, we explore the broader portfolio of use cases: sectors where hydrogen's particular strengths, or the absence of better alternatives, might yet justify its cost.

Big Numbers: Chapter 4

- **$30 per megawatt hour** - Cost of shipping energy as liquid ammonia: 60% cheaper than shipping liquid hydrogen, but...

- **500 ppm** - Concentration in air where ammonia chemically burns eyes and respiratory system. A dangerously toxic substance, requiring careful handling.

- **60–90%** - CO_2 saving from replacing fossil jet fuel with HEFA biojet, but...

- **2–3x** - The cost of HEFA fuels compared to fossil equivalents today.

- **70–98%** - CO_2 saving from replacing fossil jet fuel with e-kerosene, but...

- **4–5x** - The cost of e-fuels compared to fossil equivalents today.

PART 2

Deployment: Where Hydrogen Fits
(And Where it Slips)

CHAPTER 5

Hydrogen for Industry

Grey to Green

Naturally, when it comes to the question of where our hypothetical future supplies of clean hydrogen should be used, the place to start is where we use our existing supplies of dirty hydrogen. Almost all of the 95 Mt of hydrogen produced in 2022 was made from unabated steam reformation of natural gas (SMR), and almost all of this was consumed by the oil refining and chemical industries.[1] Oil refining was the single biggest user, gobbling up 41 Mt and releasing somewhere between 240–380 Mt of CO_2 as a result: 0.5% of global emissions. Ammonia production was next with nearly 32 Mt consumed, followed by methanol production at 16 Mt. The greenhouse gas emissions resulting from hydrogen produced for the chemicals sector totalled around 680 Mt, which was 1.2% of the global total. Clearly, replacing the grey hydrogen used by these industries with clean hydrogen is a worthy goal for reasons of climate security.

Unfortunately, swapping out grey for green is not as simple as changing the batteries in your remote control. Large chemical process plants are designed to minimise the amount of energy that is wasted, and to recoup as much value from the feedstocks as possible. This means that grey hydrogen production is commonly integrated into the overall process and shutting it down could be analogous to removing a bodily organ. Let's look at each industry individually to see why.

Oil Refining

In oil refining, hydrogen is mostly used to remove impurities and increase the amount of high value transport fuels that is recovered from crude. Around half of the hydrogen used in refineries is consumed in a process step known as hydrotreating. This is where hydrogen is mixed with fossil fuel at high temperature and pressure in the presence of a catalyst for the purpose of removing unwanted elements. Hydrogen reacts with these elements – metals, nitrogen, oxygen, and most notably sulphur, to produce molecules that are easier to separate out in a later step. This is sort of like spraying a stain remover on your clothes before running them through the washing machine. Hydrotreating helps refineries produce fuels that burn cleaner and with fewer dangerous emissions.

DOI: 10.1201/9781003361428-7

The second hydrogen hungry process is hydrocracking, which accounts for around 40% of the hydrogen used in oil refineries. In a hydrocracker unit, the heavy, gunky component of crude oil is bombarded with hydrogen at even higher temperatures and pressures. This breaks down the long hydrocarbon chains into smaller, more valuable molecules such as those that make up gasolene and jet fuel.

Both hydrotreating and hydrocracking use hydrogen for its molecular properties; as a chemical reactant. Refineries that produce excess hydrogen may also use it for its energy content, burning it to generate heat for elsewhere in the process. This would account for less than 10% of overall consumption however.

The opportunity to reduce emissions from refinery hydrogen consumption is limited because about two thirds is a byproduct of core refinery business. Obviously, it makes no sense to throw this away and buy in clean hydrogen to replace it. However, it's a different story for the third of refinery hydrogen that is purposely made to top up the byproduct supply, and this has been creeping upwards in recent years.

Crude oil supplies have been trending toward higher sulphur content over time, making refining more energy-intensive and costly. This is because the principle of 'best first' meant that we prioritised the extraction of low-sulphur crudes, which are easier to process. More sulphur in the oil means that more hydrogen is required to remove it. Given that refineries have limited space, and that it is expensive to make alterations like increasing the size of an SMR plant, the solution to rising hydrogen demand has increasingly been to buy it in. For example, the US Energy Information Administration (EIA) has reported that the American refining industry obtained 70% of its top-up hydrogen from merchant suppliers in 2022, which was up from 63% just three years earlier.

Because this piped-in hydrogen is not integrated with the wider refinery it can be made any which way, meaning the existing supplies of grey could be replaced with green. Of course, cost will be an important factor as hydrogen supply already accounts for between 10–25% of refineries' operating expenses according to energy consultancy Wood Mackenzie – and that's with hydrogen costing around $1 per kilogram. For green hydrogen to be economically viable, its production cost— currently averaging $6–7 per kilogram—will need to fall substantially, likely alongside the introduction of a carbon price that raises the cost of fossil hydrogen. Assuming the economics can be worked out, Wood Mackenzie has also estimated that green hydrogen could provide oil refining emissions reductions that top out at around 35%.

Interestingly, at a time when many companies are delaying their decisions on whether to invest in green hydrogen due to uncertainties around legislation and the willingness of buyers to pay, it is the oil refineries that make up a majority of the early movers. As of mid-2024, only five 100 MW-scale projects had received final investment decisions in Europe, which at the time was the territory with arguably the most certainty around government support. Three of these five projects involve oil companies who intend to use the green hydrogen to displace fossil-hydrogen in their existing operations. These were Shell's Hydrogen 1 and Total Energies' OranjeWind, which are both in the Netherlands, along with Shell's Refhyne 2 in Germany. It is thought that the big oil companies have an easier decision to make because they can buy their own production, which eliminates the need to seek out and lock in customers to long term contacts.

Ammonia Production

On a product weight basis, ammonia ranks high on the list of most polluting industrial products. The IEA estimates that the industry emits around 2.4 tonnes of CO_2-equivalent for every tonne of ammonia produced.[2] In 2022, this meant 410 Mt of carbon dioxide, which was nearly 0.7% of global emissions. These emissions are generated in a few different ways, so a basic understanding of ammonia production is required to appreciate why it is commonly brought up as a prime opportunity for green hydrogen.

Ammonia is produced on an industrial scale in large plants that use the century old Harber-Bosch process, which has been fine-tuned over decades to maximise the extraction of hydrogen and energy (in the form of steam) from fossil fuels. This method combines nitrogen from the air with hydrogen over a catalyst at high temperature and pressure: 400–500°C and 150–300 atmospheres. The recipe requires three parts hydrogen to one part nitrogen, with an overall reaction that is simple to understand:

$$3 \, H_2 + N_2 \rightarrow 2 \, NH_3$$

hydrogen gas + nitrogen gas → ammonia gas

An ammonia plant is really a combination of two distinct processes: one that produces hydrogen from fossil fuels, and another that synthesizes ammonia from hydrogen and nitrogen. The hydrogen production side, or 'hydrogen loop,' requires significant heat input to drive its reactions. The ammonia synthesis side, or 'ammonia loop,' generates heat through its chemical reactions, which are exothermic. The heat is captured as steam and redirected to help power the hydrogen loop. This process integration reduces the amount of fuel needed for hydrogen production, making the overall operation more energy efficient.

Natural gas is the hydrogen source for 70–80% of the world's ammonia. About two thirds of the total CO_2 emissions from ammonia production come from what's left of natural gas once the hydrogen has been extracted, as the equation below shows. A well running plant might produce 1.6–1.9 t CO_2/t NH_3, which is roughly half of what is achievable if coal or oil are used instead of gas.[3]

$$CH_4 + 2 \, H_2O \rightarrow CO_2 + 4 \, H_2$$

methane gas + steam → carbon dioxide gas + hydrogen gas

The remaining third of an ammonia plant's emissions comes from the exhaust of natural gas-fired boilers, which supply the steam needed for hydrogen production. As with all natural gas use, there are also significant upstream emissions from methane leakage during extraction and transport. These are estimated to fall in the range of 0.2–2.4 t CO_2/t NH_3.[4]

On the surface, substituting green hydrogen appears to be a simple way to eliminate emissions from ammonia production. In reality, the high degree of heat integration in existing processes makes this difficult to achieve. According to the Ammonia Energy Association, substituting more than 10–20% of a plant's hydrogen supply with green hydrogen is not possible without extensive modifications. A recent project by Spanish fertiliser manufacturer Fertiberia illustrates the point. Its existing 200,000 tonne-per-year ammonia plant introduced green hydrogen from the nearby 20 MW Puertollano electrolyser facility, which was Europe's largest at the time. However, as reported by Iberdrola—owner of the Puertollano plant—natural gas consumption was reduced by just over 10%.

Fully green ammonia production is more feasible in a purpose-built plant. Without a steam methane reformer, heating requirements are greatly reduced. Hydrogen production, the ammonia loop, and the air separation unit (which provides nitrogen) can all be powered by renewable electricity. The main challenge then becomes the intermittent nature of renewables, since an ammonia synthesis reactor needs to operate continuously. Technically, this can be addressed by building batteries and hydrogen storage tanks: making hydrogen while the sun shines and using it to keep the process running when it does not. However, this approach can result in an expensive plant, which pushes up the cost of the ammonia it produces.

The cost of green ammonia is heavily dependent on the price of electricity. As a result, the lowest production costs will be found in locations with abundant, cheap renewable energy. This relationship is illustrated in Figure 5.1, based on indicative figures from the IEA for ammonia produced from variable renewable electricity with hydrogen storage. In cheaper regions such as China and the United States, green ammonia currently costs between 1.5 and 4.5 times more than conventional fossil-based production. In regions with higher electricity prices, such as Western Europe and Japan, green ammonia would be three to six times more expensive. Fossil-based ammonia has generally been priced between $350 and $600 per tonne in recent years, suggesting that the middle to upper end of the range shown in Figure 5.1 is more realistic.

The 2030 projections assume that countries deliver on their existing climate pledges, leading to lower costs through cheaper renewable electricity and technological improvements. Under this scenario, green ammonia could be competitive with fossil ammonia in the best locations, while remaining marginal or more expensive elsewhere. In short, while green ammonia is technically feasible and offers the potential for deep emissions cuts, realising that potential requires the construction of new fully electric plants, which in turn leans heavily on the continued expansion and cost reduction of renewable power.

Unsurprisingly, given the preference of export-focused developers to move renewable energy in the form of ammonia, the largest green hydrogen facility currently under construction will also house the world's largest green ammonia plant. The NEOM green hydrogen project—located in Saudi Arabia and carrying a price tag of more than $8 billion—is targeting first operations in 2026 or early 2027. It will draw on nearly 4 GW of combined solar, wind, and energy storage to produce

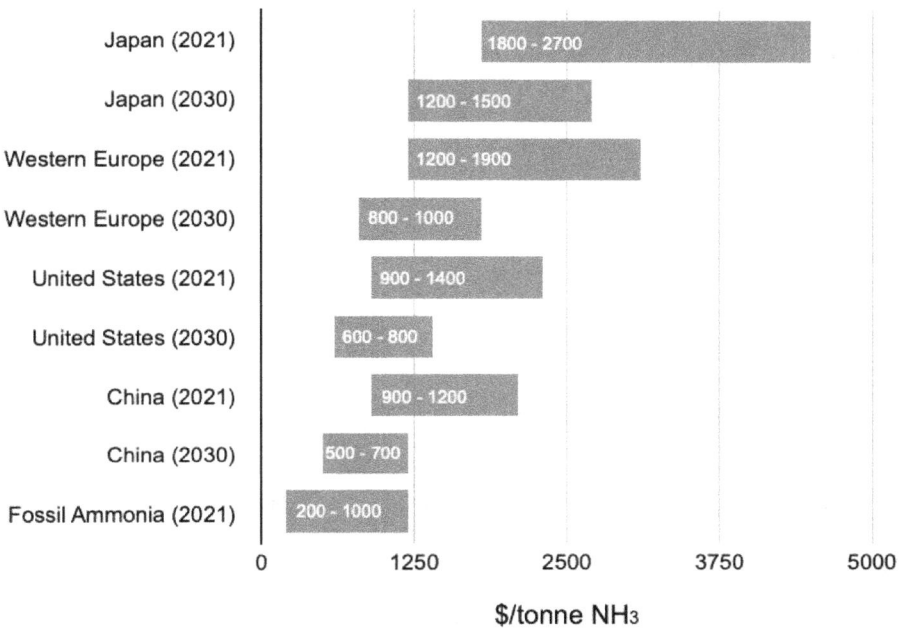

Figure 5.1. Indicative production costs for green ammonia. Source: IEA (2023).[5]

600 tonnes of green hydrogen per day, which will be converted into up to 1.2 million tonnes of green ammonia per year, according to the developers.

The ammonia is expected to be purchased by one of the NEOM project's partners, the industrial gases multinational Air Products. It is anticipated that it will be exported to Europe. However, the end use of the ammonia remains uncertain, as contracts have not yet been finalised and the market for clean hydrogen derivatives in Europe is still evolving.

Methanol Production

Methanol is the simplest molecule in the alcohol family. It's mostly used in the production of plastics but it is also a feedstock for resins, dyes, perfumes and pharmaceuticals. These applications all use the methanol molecule as a chemical building block, but it is a potential fuel as well. It is considered one of the lead contenders for decarbonizing shipping, along with ammonia.

The production of methanol also shares many similarities with ammonia. Both are based on the availability of cheap natural gas and use steam methane reformation to crack the gaseous hydrocarbons into hydrogen and carbon monoxide. The difference is that rather than getting nitrogen involved as happens in ammonia production, these compounds are simply recombined to produce methanol. This happens in the presence of a catalyst at 200–300°C and 50–100 atmospheres of pressure. The simplified overall reaction can be expressed as follows:

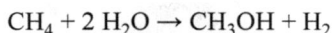

$$CH_4 + 2 H_2O \rightarrow CH_3OH + H_2$$

methane gas + steam → methanol + hydrogen gas

As with ammonia, methanol production generates greenhouse emissions from the SMR reactor, from burning fuel to run it, and from upstream gas leakage. Methanol production released over 260 Mt of carbon dioxide emissions in 2022, or around half a percent of global emissions.[6] This equates to a carbon intensity of 2.3 Mt CO_2 per tonne of methanol, which is only slightly lower than ammonia. The reason that the methanol industry has a notably smaller carbon footprint than ammonia is simply because it produces less product.

Another important trait shared with ammonia is that the heat integration of existing plants limits how much green hydrogen could be substituted for natural gas. Methanol's dependence on natural gas is even deeper than process design. While ammonia only needs the hydrogen contained in methane, methanol requires both the hydrogen and the carbon atoms. Consequently, there is limited scope for decarbonising existing methanol plants. If we're talking about new purpose-built plants using a different process however, then there are options.

So-called green methanol can be made in a couple of ways. The first is known as bio-methanol and uses the conventional SMR process with the difference being that the methane comes from a biological source rather than being extracted from the ground. This 'biomethane' or 'sustainable natural gas' is produced through anaerobic digestion, which is where microorganisms break down organic material in the absence of oxygen. A wide range of feedstocks can be used including manure and crop residues, sewage sludge from wastewater treatment plants, and waste from industrial food processing. Green hydrogen does not play a role in bio-methanol production.

The second option is to produce e-methanol, which was introduced in the previous chapter when we described electrofuels. Green hydrogen plays a leading role in this method, reacting with carbon dioxide at a temperature of 200–300°C and pressure of 50–100 atmospheres to produce methanol and water like so:

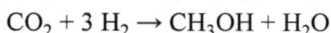

$$CO_2 + 3 H_2 \rightarrow CH_3OH + H_2O$$

carbon dioxide gas + hydrogen gas → methanol + water

The catch for e-methanol is that it requires a supply of carbon dioxide to dance with the green hydrogen. This is somewhat ironic given that the main motivation for producing e-methanol is that we have too much CO_2 floating around to begin with. The carbon dioxide must either be captured directly from the air or sourced from a waste stream with sufficiently high purity. The challenges around the high cost of direct air capture, and the limited availability of biogenic CO_2, were discussed in the previous chapter.

As with the other e-fuels, cost and carbon availability are the twin challenges. The cost of fossil methanol has typically hovered in the $100–$250 per tonne range.

Table 5.1. Estimated cost of green methanol production.

Green Methanol Route	Cost Range ($/t)
Bio-methanol	330–1000
E-methanol using by-product CO_2	800–1600
E-methanol using DAC CO_2	1100–2400

Source: IRENA/Methanol Institute (2021).[7]

As Table 5.1 shows, costs for green methanol are a lot higher. Bio-methanol, which doesn't require hydrogen is competitive in best case scenarios, but even the optimistic case for e-methanol is around 2.5 times the price of fossil-methanol. At the higher end of the range, e-methanol from waste CO_2 is nearly five times more expensive. This mirrors the cost gap seen with the Fischer–Tropsch fuels.

The economics of e-methanol do not appear to dissuade everyone however, and China in particular appears willing to throw money down. Not one, but two Inner Mongolia 100,000 tpa e-methanol plants were reported to have begun construction in late 2024.

The automaker Geely is building a plant in a region known as Alxa League, with the intention of supplying methanol for cars. China is unique in that it is dabbling in methanol as a car fuel in addition to electrification and hydrogen. Likewise, China Coal Group's plant situated near the city of Ordos plans to use 625 MW of renewable power and hopes to begin production in 2027. Both plants have said they will use CO_2 captured from nearby industries, but it is unclear what these industries are and consequently how clean the e-methanol will be.

China Coal also has approval to build a 200,000 tpa methanol plant that will use renewable electricity and CO_2 from gasified cow dung: an interesting way of wringing a little extra value from some of Inner Mongolia's more than one million cows.

Meanwhile, the first e-methanol to be certified as green under European Law was produced in early 2025. The milestone was achieved by European Energy's Kassø plant, a facility tucked near a small village in southern Denmark. Once fully ramped up, it is expected to produce around 42,000 tonnes of e-methanol per year. The hydrogen is made onsite using electrolysers powered by a 300 MW solar park, backed up by grid electricity. The CO_2 comes from a nearby biogas plant that processes agricultural waste. Prospective buyers are expected to include shipping companies and industries looking for ways to stay on the right side of Europe's tightening carbon rules.

With early e-methanol projects like Kassø coming online, and major green ammonia projects like Neom poised to join soon, it is useful to consider how this new wave of production may stack up against fossil incumbents - and against each other.

Table 5.2 provides estimates of the production costs, well-to-gate emissions (WTG), and the price of carbon abatement for our two hydrogen-hungry chemicals. The WTG emissions, which cover production but not use, are preferred to full lifecycle emissions because the later varies wildly depending on how the chemicals are used. For example, feeding ammonia to a fuel cell will produce no emissions, but

Table 5.2. Economic and environmental comparison of green ammonia and e-methanol.

	Production Cost ($/GJ)	WTG Emissions (kg CO_2/GJ)	Carbon Abatement ($/t CO_2)
fossil ammonia	24–48	100–170	--
green ammonia (2023)	50–240	near zero	14–1540
green ammonia (2030)	30–150	near zero	−130–900
fossil methanol	5–13	40–225	--
e-methanol (2020)	40–80	1–23	1170–3260
e-methanol (2030)	20–40	1–23	300–1520

Sources: NH_3 data; IEA (2024).[8] CH_3OH costs; IRENA/MI (2021).[7] CH_3OH emissions; Methanol Institute (2022).[9]

turning it into fertilizer and feeding it to crops can release substantial quantities of the powerful greenhouse gas nitrous oxide. Likewise, burning methanol will liberate CO_2 but turning it into a car taillight wont. When the use-case is simply replacing fossil hydrogen with a renewable electricity based version to produce a commodity chemical, it makes sense to stop counting emissions at the gate.

So, assuming that all the major production steps are powered by renewable electricity, green ammonia will provide near zero emissions. Methanol's emissions will not be zero, but will still be very low with a small amount of variation coming from the CO_2 source. The large variation in fossil production for both chemicals reflects the difference between making hydrogen from natural gas, which is much cleaner, and from coal, which is common in China. E-methanol can therefore offer emissions reductions in the range of 40–98% depending on the CO_2 source.

Turning now to costs and carbon abatement, we see that green ammonia today has a wide range. This is mostly due to the price of plant construction, which varies widely by location and can be very high. By contrast, the cost of producing e-methanol is determined more by the price of electricity and captured CO_2, rather than by complex plant infrastructure. Because these input prices are more predictable, the range of estimated production costs for e-methanol is narrower.

While the range of production costs is extraordinarily broad, the IEA notes that the cost of green ammonia production in 2023 was a little less than three times the average cost of fossil ammonia, implying that a figure around $100/GJ is the best representation. This translates to a carbon abatement cost in the $400–500/t range: around ten times the typical carbon price. For e-methanol, the price of avoiding emissions, which currently starts at nearly $1200/t is simply wild. Neither technology appears economically feasible based on these estimates.

The 2030 projections assume improvements in electrolyser efficiency, and continued reductions in renewable electricity costs. Under these assumptions, we see that green ammonia could easily become a commercially viable low-carbon product – in the right circumstances. In the best locations, it may even be cheaper than fossil ammonia. Methanol on the other hand, looks to remain a relatively expensive decarbonization option even when future optimism is priced in. In the most favourable of circumstances, e-methanol may be workable with government support but lower-cost production pathways like bio-methanol may ultimately prove to be a more attractive decarbonisation option.

Industrial Process Heat

The large industrial processes that create the building blocks of our material world are hungry beasts. In 2022, these behemoths consumed 20% of global energy supplies just to generate heat. Most of this heat is obtained by burning fossil fuels, which is a major contributor to industry's greenhouse gas emissions. Estimated at nine gigatonnes of CO_2, these represent a quarter of global energy-related emissions.[6] The sheer scale of energy use makes process heat a key target for decarbonization efforts. However, each industrial sub sector has different requirements and challenges when it comes to subbing out fossil-energy.

Firstly, there is a wide range of temperature requirements within the industrial sector. According to the IEA, 44% of process heat is at a low temperature, which is defined as less than 200°C. At the opposite end of the scale, 50% of process heat is considered high temperature, meaning hotter than 400°C. This leaves around 6% of processes somewhere in between.

If it is viable, electric heating is preferred to burning fuel due to its superior energy efficiency. Industrial heat pumps are the most efficient option and are improving all the time. Heat pumps are extraordinarily efficient because they don't generate heat—they move it.

Think of a heat pump as a train running between two platforms on an oval track, picking up a load on one side and dropping it off on the other. In this case, the 'train' is a fluid with a very low boiling point—perhaps –50°C—circulating through coiled tubing like that found on the back of a refrigerator. As it moves outside, the fluid absorbs heat by evaporating. This heat can be drawn from the air, ground, or water. When the vapour reaches the indoor unit, it condenses back into liquid, releasing the captured heat. The fluid is circulated by an electric compressor, and industrial heat pumps can provide more than three units of heat for every unit of electricity they consume when the temperature difference between source and destination is between 30–50°C.[10]

Commercial units that can provide heat up to 100°C are already well established and widely used. The latest models can provide heat up to 140°C, with demonstration-stage technology operating at 160°C. Units that can achieve 200°C are currently in development.

Industrial heat pumps could theoretically provide almost half of the process heat required by industry. For the other half, however – those requiring temperatures above 200°C – heat pumps just aren't up to the task.

A range of direct electrical heating methods are available and some mirror technology that is found in the home. These include microwave, and induction heating, along with resistance heating, which is where electricity runs through a hot wire like in a hair dryer. There are also more exotic electrical methods such as electric arc, which is like controlled lighting, and plasma, which is a bit like heating with a tiny star. However, for very high temperature processes – those in excess of 500°C – direct electrical methods are often inadequate or unable to fit in with the existing

Table 5.3. Process heat requirement and CO_2 emissions for notable industries.

Industry	Fraction of Process Heat Greater than 500°C[1]	Global CO_2 Emissions in 2022 (Gt)[2]	Fraction of Industrial Emissions in 2022
Iron and Steel	60%	2.6	29%
Cement	80%	2.4	27%
Chemicals	25%	1	10%
Oil Refining	12%	0.5	5%
Aluminium Smelting	95%	0.3	3%
Pulp and Paper	6%	0.2	2%

Sources: 1. Deloitte (2022).[11] 2. IEA (2023).[10]

process. This is where hydrogen comes in. Capable of generating temperatures up to 2100°C when burned, hydrogen can theoretically provide heat for any of the common high temperature industries.

Table 5.3 lists the major greenhouse emitting industrial sectors, which are obviously the main focus for industrial decarbonisation, along with an indication of how much the processes rely on high temperature heat. Together, these half dozen sub-sectors release more than three quarters of industrial greenhouse gases. Not all of these emissions are from burning fuel however. The decarbonisation potential of hydrogen depends heavily on the nuances of each process and how the emissions are generated. For example, oil refining, which was discussed earlier, generates two thirds of its process heat from by-products and therefore has limited use for hydrogen as a fuel.[11] Let's see if we can't find a use elsewhere.

Aluminium Smelting

Aluminium metal is produced electrolytically; similar to green hydrogen. In this case, the raw material – an aluminium oxide known as alumina – is dissolved in a 1000°C chemical 'bath' and zapped with electricity to separate the aluminium and oxygen atoms. Most of carbon dioxide emitted from aluminium smelting is born from a reaction between the oxygen atoms and consumable carbon anodes used to carry electrical current from the bath. Furthermore, the process heat is generated internally, rather than from a secondary fuel. Electric current passing through the molten bath generates resistive heat – much like the elements in an electric toaster.

Substituting hydrogen for electricity in this process is not technically feasible. Now, baking the consumable carbon anodes is sometimes fuelled with natural gas, which could be replaced with hydrogen. However, this represents only a few percent of a smelter's energy consumption and emissions. The potential for hydrogen to decarbonise aluminium smelting is therefore very limited.

Pulp and Paper

Approximately 95% of the pulp and paper industry's emissions come from feeding its boilers. If the boilers are the heart of the operation, then the steam they produce is the lifeblood, coursing through the plant to soften wood chips, separate cellulose fibres, and dry the pulp into sheets of paper.

Over 75% of the heat is in the low temperature range and could be serviced by low-cost solutions: industrial heat-pumps or burning biomass. Furthermore, pulp and paper is a bit like refining in that it can lean on by-products and waste streams for its energy requirements. Around half of the boiler energy is derived from 'black liquor' – a byproduct of the pulping process. Buying in a fuel to replace this and then having to dispose of it would raise costs and does not make for an intelligent plan. It appears that there is little room for hydrogen in the pulp and paper industry either.

Chemicals

We've already discussed ammonia and methanol – the chemical heavyweights of hydrogen demand. Further down the leaderboard is a cluster of high value chemicals, mostly used as feedstock for other products. This illustrious lineup includes names like ethylene, propylene, benzene, toluene and mixed xylenes. The combined emissions from production of these chemicals are similar to that of the aluminium and pulp and paper industries.

Ethylene is the high value chemical that is made in the largest quantities. It is the building block of polyethylene – a common plastic used in many everyday items like bags, bottles, and containers. The major source of emissions in ethylene production is the fossil fuel that is burned to power the steam-cracking furnaces, which produces around 1.2 tonnes of CO_2 for every tonne of ethylene. Steam-cracking is a process where light factions of natural gas and oil are heated to temperatures in the 750–850°C range, until they split into even smaller molecules.

The industrial gases firm Linde recently announced that it will spend over $2 billion on a blue hydrogen plant in Canada to supply heating fuel for ethylene production. In an environment where most clean hydrogen projects struggle to gain the green light, those that do provide an indication of what the low hanging fruit applications are. Putting aside questions around the climate credentials of blue hydrogen, this development shows that there clearly is a case for hydrogen heating in high value chemical processes.

Cement

Combining for over half of all industrial emissions are the construction industry staples: steel and cement. Both require very high temperatures and could use hydrogen as a fuel. Steel-making is a fascinating use case because it can also use hydrogen as a chemical reactant, and it's important enough to warrant its own section, which is coming up shortly.

Cement making on the other hand, is strictly looking at hydrogen as a fuel for its kilns.

The cement kiln is a rotating metal cylinder about the size of a subway train. It is angled slightly so that the material fed in at the top slowly bumbles down its length, all the while baking in the heat of an intense fire situated at the bottom of the cylinder. They run a temperature of 1200–1500°C, which is hotter than molten lava. Easily exceeding the temperature needed to incinerate hazardous waste, with planning they can be configured to consume just about anything.

Cement is made by passing a mixture of limestone and other materials, such as clay or shale, through the kiln: a process called calcination. This drives off carbon dioxide and produces a material known as 'clinker' – a hard, rocky material which is ground into a fine powder and mixed with additives to form the final product.

Cement-making produces carbon dioxide in two ways. The first is the combustion of fuel to provide process heat. The second is the calcining reaction that transforms the raw limestone into lime, which is a key ingredient in clinker. The reaction can be expressed as:

$$CaCO_3 \rightarrow CaO + CO_2$$

solid calcium carbonate (limestone) → solid calcium oxide (lime) + carbon dioxide gas

The calcining reaction produces around 0.5 kilograms of CO_2 per kilogram of cement. This is the case no matter what fuel is used to provide the process heat. If calcium carbonite is the raw ingredient, carbon dioxide will be a product.

The reaction is endothermic; requiring heat energy which is provided by the fire. The emissions from burning fossil fuels to produce the fire typically result in an additional 0.3–0.4 kilograms of CO_2 per kilogram of cement. Overall, cement emissions are roughly 60% calcining reaction and 40% fuel combustion. This means that using low carbon hydrogen as kiln fuel can reduce cement emissions by no more than 40%. However, given the scale of cement production, even 40% reduction industry wide would mean a 1% drop in global emissions.

Hydrogen has been blended into the fuel feed in several trial projects in Europe and the UK. As with other applications, substituting hydrogen for fossil fuels also raises technical challenges around its flame characteristics, which limits the amount that can be substituted with existing equipment. However, the key barrier to using hydrogen as fuel will be its price and availability. Furthermore, there are a host of competing solutions for decarbonising cement production such as carbon capture, alternative cement blends and chemistries, and burning biomass or waste streams as fuel.

In its decarbonisation roadmap, the European Cement Association (Cembureau) anticipates that hydrogen and electrification will contribute only 4% to the aspirational reduction in clinker's carbon intensity by 2050.[12] So, while hydrogen may play a role as heating fuel in some cement plants, this will be far from a universal solution.

Steel

Steel is the figurative backbone of modern civilization, and often the literal skeleton of our material world. From towering skyscrapers and expansive bridges to cars, ships, and railways, steel lends its strength and durability to our critical infrastructure and is essential in manufacturing machinery, appliances, and tools. Steel's versatility and affordability make it indispensable, and we are currently producing nearly 1.9 billion tons of the stuff each year. Unfortunately, steel production has historically relied heavily on fossil fuel, and in particular – coal. Consequently, steel production in recent years has contributed around 2.7 billion tonnes of carbon dioxide annually, which is 7% of all energy related emissions.[10] Decarbonising steel production is a necessity.

There are a few different methods by which steel is produced. The most widely used by far, accounting for around 70% of steel production and 85% of the industry's emissions, is the two step Blast Furnace – Basic Oxygen Furnace route (BF-BOF).[13] Resembling a 30 to 40 floor high lava lamp, the blast furnace's role is to transform pellets of iron ore to pure iron. The iron ore is fed into the top of the furnace along with limestone and a pure form of carbon known as 'coke'. Air is preheated to around 1600°C and blasted in from the bottom of the furnace (hence its name). This intensely hot air ignites the coke, which generates even more heat and produces carbon monoxide which helps reduce the ore via the simplified reaction:

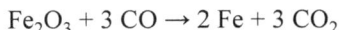

$$Fe_2O_3 + 3\ CO \rightarrow 2\ Fe + 3\ CO_2$$

(hematite/iron oxide) + carbon monoxide → iron + carbon dioxide

The limestone's role is to remove impurities from the iron by reacting with them to form a material known as slag—a useful byproduct with numerous applications in construction and industry. The molten iron is 'tapped' from the bottom of the furnace and, once solidified, is a relatively soft form known as pig iron. While pig iron is sometimes used as is, it is more commonly sent directly to a basic oxygen furnace (BOF) for further processing into steel, which is harder and stronger.

The basic oxygen furnace can be imagined as a giant wizard's cauldron—bubbling and churning as ingredients are hurled in. Like magic potions, steel is made in batches. The 'cauldron' is first filled with molten iron, with scrap steel often added to help control the temperature. What makes steel different from iron is the presence of a small amount of carbon—typically 0.05 to 2% by weight. Like chocolate chips in a cookie, these carbon atoms are scattered throughout the iron, forming a 'solid solution'. The carbon atoms lock the iron atoms in place preventing them from slipping past one another. That's what gives steel its strength.

Molten iron leaving the blast furnace already contains a high level of carbon (3 to 4%) left over from the coke. This is actually too much for making steel. To reduce it, high-purity oxygen is blown into the molten metal through a specially designed, water-cooled lance, inserted like a giant baster. The oxygen reacts with the excess carbon to form carbon monoxide and carbon dioxide, which escape as gas. These reactions also release intense heat, cleverly eliminating the need for additional fuel. After about 20 minutes of 'basting', the carbon content drops to the desired level, and the furnace is tipped to pour out the newly formed molten steel.

Both the blast furnace and the basic oxygen furnace release carbon dioxide. The blast furnace is responsible for around 70–85% of the overall emissions from the two-step steel-making process however.

A lower-carbon method of steel-making commonly employed today is the electric arc furnace (EAF). Electric arc furnaces have a similar role to the BOF in that they are used for producing steel from metal, rather than metal from ore. Currently used for almost 30% of steel production, the electric arc furnace works by passing powerful electric currents through graphite electrodes to generate what is essentially controlled lightning. This electric arc jumps through an air gap between the electrodes, generating intense heat that can reach up to 3500°C – enough to melt metal with ease. The molten bath itself, however, typically sits at a balmy 1800°C.

Electric arc furnaces are used to make both primary and secondary steel. Primary steel is that which originates from iron ore. EAF can be integrated with virgin iron production as an alternative to the basic oxygen furnace. More commonly however, electric arc furnaces are fed with scrap metal, the product of which is termed secondary steel, which essentially means recycled.

Recycling usually provides substantial energy and emissions savings, compared to producing a material from scratch. This is certainly the case for steel. According to the World Steel Association, using global-average grid electricity, the scrap-based EAF route emits 0.68 kilograms of CO_2 per kilogram of steel. Even when powered by a dirty fossil-heavy grid, scrap-EAF produces less than a third of the emissions from the BF-BOF route, which is 2.33 kilograms CO_2 per kilogram steel.[14]

Unfortunately, secondary steel is tied to the availability of iron-based scrap metal which limits the scale of production. Another alternative that is of increasing interest for decarbonizing primary steel is to feed an EAF with directly reduced iron (DRI). Direct reduction operates at temperatures below the melting point of iron and uses an oxygen-hungry reducing agent to strip out the oxygen contained in iron ore – leaving behind a porous 'sponge iron'.

Usually, the DRI reducing agent is natural gas, although coal can also be used. These fossil fuels are reformed into the powerful reductants carbon monoxide and hydrogen. These are the agents that directly react with the ore. A simplified overall reaction demonstrating fossil-DRI is as follows:

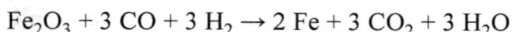

$$Fe_2O_3 + 3\ CO + 3\ H_2 \rightarrow 2\ Fe + 3\ CO_2 + 3\ H_2O$$

hematite/iron oxide + carbon monoxide + hydrogen → iron + carbon dioxide + water

At 1.37 kilogram CO_2 per kilogram steel, the fossil-DRI-EAF method sits squarely between BF-BOF and Scrap-EAF in terms of carbon intensity. This is currently a minority method, accounting for around 7% of steel production.

Now, an astute reader may have noticed that hydrogen is one of the DRI reducing agents, implying that this method could be carbon free. This is precisely why hydrogen is of great interest for steel decarbonization. With hydrogen as the sole reducing agent, iron may be produced from ore via the following simplified reaction:

$$Fe_2O_3 + 3\ H_2 \rightarrow 2\ Fe + 3\ H_2O$$

hematite/iron oxide + hydrogen gas → iron + water

When both the hydrogen and the electricity for the furnace are green, H_2-DRI-EAF steel is thought to reduce carbon emissions by 95% compared to fossil-DRI-EAF.[14] Decarbonisation doesn't get much deeper than that.

Table 5.4. Comparison of steel production methods.

Process Route	Fraction of Global Steel Production[1]	Avg GHG Emissions (t CO_2/t steel)[1]	Avg Energy Consumption (kWh/t steel)[2]	Energy Carriers Used[3]
Primary Steel				
BF-BOF	70%	2.3	6000	90% coal 7% electricity 3% natural gas
Fossil DRI-EAF	7%	1.4	4750	50% electricity 38% natural gas 12% coal
H_2-DRI-EAF	NA	< 0.1	3000	electricity; renewable
Secondary Steel				
Scrap-EAF	23%	0.7	600	electricity; world avg grid
Scrap-EAF	--	< 0.1	600	electricity; renewable

Sources: 1. WSA (2024).[14] 2. IEEFA (2022).[15] (H_2-DRI assumes 60 kg H_2/t steel and 50 kWh/kg H_2). 3. WSA (n.d.).[16]

The energy and carbon intensities of green H_2-DRI-EAF are compared with current steel production methods in Table 5.4. We can see from these numbers that recycling steel in an EAF is by far the least energy intensive method, and when powered by renewables has a very small carbon footprint. For primary production, a similar carbon intensity is available through Green H_2-DRI-EAF. However, this requires five times the energy, which shows why we should recycle as much steel as we can.

Green steel is not yet commercially produced but there are a handful of developments that could see this change in the coming years. In 2021, Reuters reported that the first industrial-scale DRI using hydrogen had been achieved by Sweden's Hydrogen Breakthrough Ironmaking Technology project (HYBRIT). That same year, they made the world's first delivery of fossil-fuel free steel, which was received by Volvo. HYBRIT is a joint venture between steelmaker SSAB, state-owned power company Vattenfall, and Europe's largest ore producer LKAB. Having completed its pilot project phase, HYBRIT is now planning to build a 1.3 Mt per year commercial-scale hydrogen DRI demonstration plant in the Swedish town of Gällivare.

Stegra, another Swedish company, is currently building an integrated H_2-DRI-EAF plant on a massive 270 hectare site near the town of Boden. In addition to iron and steel making units, this site will incorporate 700 MW of electrolysers, which Stegra says will be the largest hydrogen plant in Europe when completed.

They hope to produce 2.5 Mt per year of green steel by 2025, expanding to 5 Mt by 2030. HYBRIT also hopes to have its plant running by 2025, so it looks very likely that the world's first commercial-scale H_2-DRI-EAF derived steel will come from Sweden.

Several major steelmakers—mostly in Europe—plan to replace existing blast furnaces with DRI units. However, they do not intend to produce their own hydrogen. Instead, they will begin operations using natural gas, with the option to switch to hydrogen if it becomes available in sufficient volumes and at a viable price. According to *Hydrogen Insight*, Axel Eggert, Director of the European Steel Association (Eurofer), has stated that hydrogen would need to cost around $2–3 per kilogram for hydrogen-based DRI to be competitive. With current European prices roughly three times higher, it remains uncertain when—or if—these plants will actually make the switch.

There are also a number of pilot-scale projects around the world that will produce green hydrogen and use it for DRI. These are generally in locations with access to quality renewable energy resources, such as Australia, Namibia, and the UAE. Interestingly, Stegra's Chief Growth Officer Kajsa Ryttberg-Wallgren, has revealed that simply having access to good solar or wind power does not make a location favourable to green steel production. *Hydrogen Insight* has reported her saying that in order to minimise the cost of the plant (including power storage), electrolysers should be running 90% of the time, which requires a continuous power supply. Therefore, a favourable location requires clean baseload power, which she notes is not the case in Australia.

Another key insight from Stegra's early experience is the cost of H_2-DRI-EAF steel. The company has reportedly pre-sold about 60% of its phase one output to customers including Porsche, Scania, and IKEA. *Hydrogen Insight* reported that the contracts are priced at the market rate for steel at the time of delivery, plus a 20–30% premium. For context, hot-rolled steel coil—the industry's most widely traded product—typically sells for $600 to $900 per tonne. Given the steel sector's notoriously thin margins, producers cannot absorb this markup themselves. Buyers must therefore pay a green premium of roughly $100 to $300 per tonne if they want zero-carbon steel.

Some market observers believe that it will be the willingness of customers to pay the green premium that drives green steel production. This is consistent with the makeup of Stegra's early buyers. Large companies that trade on their environmental credentials, like Ikea obviously see value in being able to provide a low-carbon offering. Automakers in particular are well placed to pay a bit more for steel given the difference between this cost and the price they achieve for selling a car. The World Steel Association has noted that the average car contains around 900 kg of steel The green premium would therefore add no more than a few hundred dollars per car. A laughable sum compared to the price tag of a new Porsche. However, the automotive sector consumes only 12% of steel production.[14] The biggest user of steel by far is construction. Buildings and infrastructure consume a little more than half of all the steel that is produced. With its countless components and massive material requirements, construction is especially vulnerable to death by a thousand cuts if costs aren't tightly managed. This industry is much less able to absorb a 20–30% price increase and as a result it is not expected to be an early buyer of green steel.

An option for managing the price of green steel in some locations is to split the DRI and EAF processes geographically. As we have seen, it is not easy to move hydrogen around efficiently, but shipping iron ingots is fairly straightforward. For this reason, there is wisdom in producing iron in locations where green hydrogen can be made cheaply and then shipping the iron to demand centres where hydrogen is expensive. This situation is actually not so different to the early days of the steel industry, where the major production centres of the UK, US, and Germany were in regions endowed with both iron ore and the coal to process it.

The EAF steelmaking step, which requires only electricity and does not produce much in the way of direct emissions, can be located more or less anywhere. It would ideally also use renewable electricity of course, but this is less critical because decarbonising iron production is a much higher priority given that it is responsible for 70–85% of steel's emissions.

Sensible as this approach is from an energy-efficiency and cost perspective, there are realpolitik constraints. Firstly, countries may be reluctant to shut down their ironmaking capacity because they see steel as a strategic material that is important for national security. Secondly, explaining that the jobs provided by a local industry are probably going to disappear for environmental reasons is a tough sell for politicians. That sort of message tends to get one kicked out of office. These pressures mean that some countries, particularly in Central Europe, may still attempt the suboptimal strategy of importing expensive hydrogen to decarbonise their legacy steel industry.

A further challenge to greening our steel is the reality that doing so will not come cheap. Building new mills, or renovating old ones, can easily top a billion dollars. According to Shell and Deloitte, decarbonising the global steel industry will require around $800 billion in investment by 2050.[17] That is approximately the gross domestic product of Poland; the ninth biggest economy in Europe. Much of this cost will no doubt be met by the taxpayer in the form of government grants and subsidies. However, given the importance of steel and magnitude of emissions from conventional production, it is an investment that needs to be made. Consequently, analysts are increasingly placing steel near, if not at the top, of the best use cases for green hydrogen.

Big Numbers: Chapter 5

- **3** - The number of processes that account for 94% of global hydrogen demand: oil refining (43%), ammonia (34%), methanol (17%).

- **35%** - Estimated ceiling on CO_2 reductions in oil refining from green hydrogen substitution, as two-thirds of both the feedstock hydrogen and the energy for process heat come from internal by-products.

- **3x** - Cost of green ammonia production compared to the natural gas method. Substantial reductions in equipment costs are required to make it competitive.

- **4x** - Minium cost of e-methanol relative to methanol produced from natural gas. The need for biogenic CO_2 inputs will likely keep costs high.

- **20%** - Share of global energy used for industrial process heat. Most major processes are poorly suited to green hydrogen substitutions; cement and some chemical manufacturing are exceptions.

- **95%** - Potential CO_2 reduction from replacing blast furnaces with hydrogen-based DRI in steelmaking. The 20–30% premium may be acceptable in high value industries, but not for the bulk of steel demand.

CHAPTER 6

Hydrogen for Heat and Power

Domestic Heat

About one and a half million years ago, give or take half a million years, early humans discovered how to control fire. This allowed them to cook their food, making it easier to digest and more nutritious, and to stay warm—opening the door to life in colder regions.

While the pace of technological change today is head spinning, these most fundamental uses of energy are just as vital as they were back then. According to the IEA, about 15% of global energy goes towards heating our buildings, with a 30:70 split between warming our water and our living spaces.[1] All this heat is also responsible for around 4 Gt of CO_2 per year: 10% of global emissions when including electricity generation.

Domestic heating therefore represents a massive decarbonisation opportunity. Table 6.1 breaks down how the world heats its buildings. We can see that the largest share by some distance goes to natural gas. This is particularly true of Europe and North America, where households are between 45–50% reliant on natural gas heating.[3] In the UK, a whopping 85% of homes use gas.[2] Globally, the heating of buildings represents a sixth of natural gas consumption. In Europe, this number is closer to a third.

To service buildings, natural gas is generally burned in a furnace or boiler, which heats either air or water, depending on the system design. Warm air is circulated through ducts, while heated water flows through radiators or underfloor piping. Natural gas systems have traditionally been favoured for reasons of efficiency and low cost. However, the greenhouse gas emissions associated with burning natural gas necessitates a move away from its use and this imperative was only accelerated when Russia's invasion of Ukraine sent gas prices through the roof – particularly in

Table 6.1. Energy sources for building heat in 2021.

Natural Gas	Oil	Electricity (resistive)	District Heating	Electricity (heat pumps)	Biomass and Coal
42%	15%	15%	11%	10%	7%

Source: IEA (2022).[3]

DOI: 10.1201/9781003361428-8

Europe. Consequently, low-carbon hydrogen has been proposed as a replacement for natural gas heating.

There is no technical reason why hydrogen cannot be substituted for natural gas, although the differences in gas properties would require boilers and furnaces to be changed, along with some network equipment. Proponents of hydrogen heating—most commonly the companies that own and operate the existing gas resources and networks – often paint hydrogen as a 'drop-in' solution, that would allow gas users to continue as they have been. This low-cost, low-disruption narrative has been attractive to policy makers because it appears to offer a net-zero transition that utilises existing infrastructure and doesn't require consumers (and voters) to do anything. Consequently, several small-scale hydrogen heating trials have been run in countries including Canada, Germany, Japan, and the Netherlands.

In fact, the Dutch village of Stad aan't Haringvliet, with a population of around 1500 people, voted in 2023 to become the first municipality to switch its gas supply to green hydrogen. Residents will receive new hydrogen boilers and piping free of charge, along with new (electric) cooktops. The project leaders intend to source the hydrogen from a nearby wind farm and hoped to make the switch in 2025, while acknowledging that this may not actually happen before 2030.

Trials like this are presented as proof of concept. However, just because something can be done doesn't mean that it should. As we have seen, hydrogen may have a role in industrial settings where very high temperatures are essential, but that doesn't mean it's a good fit for homes and businesses. Keeping an apartment at 21°C for a few months a year is not the same thing as continuously running a cement kiln at 1500°C – and the optimal solution isn't the same either.

Cost and Efficiency

A 2024 meta study[4] reviewed 54 independent research programmes on hydrogen heating—'independent' meaning they were not funded by gas companies or network operators. The composition of these studies reveal where the idea has gained the most interest. The UK led with 14 investigations, followed by Germany with nine. Thirteen were global in scope. Despite the diversity in target locations and research methodology, the findings were consistent. The meta study concluded that:

> "The scientific evidence does not suggest a major role for hydrogen heating in cost-optimal pathways, and indicates higher system and consumer costs."

Nearly half of these studies focused on identifying the mix of heating technologies that provided the lowest cost for a given region. The median result saw hydrogen contributing a lowly 1%. So, while niche uses may exist, hydrogen heating is broadly too expensive and inefficient compared to the alternatives. And what is the best alternative? Heat pumps.

The IEA's net-zero scenario envisions heat pumps rising from providing 10% of building heat today to 55% by 2050. They have stated that the annual household savings when switching from a natural gas boiler to a heat pump range from $300 in the United States to $900 in Europe: 2%–6% of a low-earning household's income.[3] Furthermore, the switch reduces greenhouse emissions by 20% when powered by

Table 6.2. Energy efficiency comparison: electric heat pump and green hydrogen boiler.

Electric Heat Pump	Step Efficiency	Green Hydrogen Boiler	Step Efficiency
1. Grid Transmission	90%	1. Grid Transmission	90%
2. Ambient Heat Addition	300%	2. Electrolytic Hydrogen	75%
		3. Gas Distribution	95%
		4. Distribution Leakage	99%
		5. Boiler	90%
Overall Efficiency	*270%*	*Overall Efficiency*	*57%*
Green electricity required for 100 kWh of heat	*37 kWh*	*Green electricity required for 100 kWh of heat*	*175 kWh*

fossil electricity, or 80% if the electricity is clean. Heat pumps are a weapon against both climate change and energy poverty.

Recalling the volume problem—that it takes 3 to 5 times more hydrogen to match the energy of natural gas—if heat pumps beat gas on cost and efficiency, they wipe the floor with hydrogen. The difference is a combination of the inefficiency of the green hydrogen supply chain, and the near-miraculous efficiency achievable through electric heat pumps. Table 6.2 compares these two technologies using representative efficiencies for each step on the path from green electricity generation to heat in a building. The real-world efficiency of each step will have some variation of course, but the end result will be much the same.

In this example, hydrogen requires 4.7 times more green electricity to provide the same amount of heat as the heat pump. This figure is on the lower end of the range, as it is generally accepted that green hydrogen heating requires between five and six times more electricity than heat pumps.[5] That means five time the land, the windmills, the solar panels, just to do the same job.

We can see from the table that the most energy intensive step in the green hydrogen route is electrolysis. If we assume that carbon capture and storage operation worked effectively, does blue hydrogen fare any better? After all, we wouldn't need to use all that electricity to make it. In this case, we compare blue hydrogen heating with the alternative of generating electricity to run a heat pump via a natural gas turbine with CCS. Modern combined cycle gas turbines (CCGT) produce power from burning natural gas as well as from the waste heat, and can achieve efficiencies of 50 to 60%.

Table 6.3 shows that the efficiency of the blue hydrogen boiler route is a little over 40%, which is about the same as delivering electricity to a building from a CCGT with CCS. The difference is the 'free' energy that a heat pump provides—and that just can't be beat. The overall result is that blue hydrogen heating requires nearly three times the natural gas to deliver the same amount of heat as the electric route.

That may be good news for the company selling the gas, but it's terrible for the consumer. The meta study referenced earlier also included 13 investigations into the cost of hydrogen heating relative to electric alternatives. All 13 found hydrogen to be more expensive, with an 86% premium being the median result. This point was

Table 6.3. Energy efficiency of a blue hydrogen boiler vs heat pump powered by a natural gas turbine with CCS.

Electric Heat Pump; Gas Turbine Power and CCS	Step Efficiency	Blue Hydrogen Boiler	Step Efficiency
1. Gas Transmission	97%	1. Gas Transmission	97%
2. Upstream Leakage	95%	2. Upstream Leakage	95%
3. CCGT Generation	55%	3. SMR Hydrogen	65%
4. CCS	85%	4. CCS	85%
5. Grid Transmission	90%	5. Gas Distribution	95%
6. Ambient Heat Addition	300%	6. Distribution Leakage	99%
		7. Boiler	90%
Overall Efficiency	*122%*	*Overall Efficiency*	*43%*

colourfully made by Greg Jackson, the CEO of Octopus Energy in the UK, when he referred to hydrogen heating as "flushing the toilet with champagne."

Heat Pump Criticism

Given the existential threat that they pose to the prospect of hydrogen heating, a number of arguments against heat pumps have been made by its proponents.

The first is that heat pumps are unsuitable for very cold climates. It's true that their efficiency drops as the temperature gap between the source and destination grows, but some heat pumps are engineered specifically for very cold conditions. Today, 60% of buildings in Norway have heat pumps, along with 40% in Finland and Sweden. Winter temperatures in all three hover around freezing—clear evidence that heat pumps can handle the cold.

A second claim is that widespread adoption would increase electricity demand and strain power grids, but there's little substance behind this. In 2021, the CEOs of Europe's largest utilities—covering Germany, France, Italy, and Spain—signed a letter to the European Commission declaring that electrified heating is a "reliable solution towards a zero-emission Europe." They also assured the EC that "the lights will stay on with 50 million heat pumps."

What's more, concerns about costly grid upgrades also ignore the fact that these upgrades are needed anyway to accommodate the energy transition. Also, switching to hydrogen isn't cheap either. In 2023, the consultancy Arup estimated that converting the UK's gas network to hydrogen would cost between $60 and $90 billion.[6] This may be compared to the $75 billion the UK's National Grid operator recommends for electricity grid upgrades to support a decarbonised system.[7]

Other criticisms of heat pumps include higher upfront costs compared to gas boilers – which can be addressed through government support – and that outdoor units can be difficult to retrofit in buildings with limited space. As with any technology, there will be cases where installation is tricky. But from a policy standpoint, it's increasingly difficult to ignore the consensus among researchers and energy experts that, in the words of the IEA, "heat pumps, powered by low-emissions electricity, are the central technology in the global transition to secure and sustainable heating."[3]

Climate, Health, and Safety

Although the efficiency shortfall is reason enough to cast aside the prospect of using hydrogen, at least for those interested in systems engineering, the physical properties of hydrogen also raise health and safety concerns that may be more meaningful on an individual level.

In Chapter 1, we described how hydrogen is likely to be around three times more leaky than natural gas under similar conditions. Assuming the same sized pipes are used, replacing natural gas with hydrogen would require higher pressure and flow rates to compensate for the lower energy density. This could further exacerbate its already higher tendency to leak.

Hydrogen's global warming potential is about a third that of methane, so we might expect the overall climate impact of 'fugitive' emissions to balance out. But even this would undermine the argument that hydrogen heating is benign for the climate.

Evidence for these concerns was provided by a recent study that looked at leakage rates from new home appliances for a gas blend of 20% hydrogen with 80% methane.[8] It found that emissions from gas boilers increased by 44% while emissions from gas cookers doubled. This cancelled out the modest 7% climate benefit – the theoretical savings from a 20% hydrogen blend. If the gas feed was 100% hydrogen or if older appliances were used, we would expect leakage rates to be higher still.

Hydrogen's leakiness in combination with its higher flammability also represents an elevated safety concern for residents. In a study commissioned by the UK government, the engineering consulting firm Arup concluded that replacing the UK's natural gas supply with hydrogen would result in four times more home explosions per year, with four times the casualties.[9] They also suggested this could be reduced to 'only' three times more if safety valves were added to the network and inside homes. The UK currently experiences around a dozen gas-related home explosions per year.

Switching to hydrogen in homes may also be detrimental to human health, as it could lead to higher concentrations of NOx – a known contributors to respiratory problems. This is less of a concern for boilers, which can be fitted with catalytic converters that remove NOx. However, this solution doesn't work for gas cooktops and fireplaces because they are open to the room.

In another UK government study, the Frazer-Nash consultancy assessed the suitability of existing domestic appliances for hydrogen and reported that hydrogen burners might run hotter than natural gas equivalents, leading to higher NOx emissions.[10] Studies on NOx emissions from blends of hydrogen and natural gas have not provided consistent results, although much of the data indicates that increasing the amount of hydrogen in the blend leads to increased NOx.[11]

There is really no reason to take these risks. As with heating, electrification of cooking outperforms hydrogen on both efficiency and safety. While induction cooktops may be more expensive upfront, they are widely considered the gold standard: requiring three to four times less energy than burning hydrogen, and delivering heat that is safer, cleaner, and more responsive.

A Case Study in Industry Resistance: Hydrogen Heating in the UK

The UK, where around 85% of homes are connected to gas, offers a revealing case study in how an entrenched industry responds when decarbonisation threatens its business model. Towards the end of 2022, a public-focused campaign cheerfully named *Hello Hydrogen* was launched, funded by British Gas, all four UK gas network operators, and several gas boiler manufacturers. The campaign aimed to "raise awareness of the exciting potential for hydrogen heating homes across the country."

This seems innocent enough but its messaging overstated the likelihood that hydrogen heating would be rolled out. The first thing a visitor to Hello Hydrogen's website will see in big bold font is "The future of home heating: hydrogen is the gas that *will* play an *essential* part in Britain's future energy supply, providing locally produced, low-carbon heat and power to millions of homes." The emphasis has been added.

In reality, the UK government's 2021 Heat and Buildings Strategy aimed for 600,000 new heat pump installations per year by 2028, while merely "developing the evidence base" to inform a decision on the future role of hydrogen heating.[12] At no point has hydrogen heating in the UK been a sure thing.

This 'fake it till you make it' strategy fell afoul of the Competition and Markets Authority (CMA), the UK's independent consumer protection body. An inquiry into the green heating sector found that "some large businesses were not being clear or upfront about the uncertainty of hydrogen deployment and inaccurately describing hydrogen rollout in definitive terms."[13]

The CMA also took issue with erroneous statements to the effect that hydrogen is a "zero-carbon fuel source" and that "water is the only by-product of hydrogen production". They concluded that some boilermakers were greenwashing, which is the practice of making something seem more environmentally friendly then it really is. In this case, manufacturers were advertising boilers as "hydrogen-blend ready", to imply they were more climate friendly and future-proof that their competitors.

In fact, no burner can operate interchangeably with 100% natural gas and 100% hydrogen due to major difference in combustion characteristics. Moreover, existing standards already required boilers to function with a 23% hydrogen blend, meaning that the hydrogen ready boilers were functionally no different to other models. As a result, manufacturers were subsequently banned from using the term.

If that were the end of it, one might dismiss this as routine corporate spin, but the following is much more damning. In 2021 – around the same time the government launched its new heat pump focused heating strategy – the Energy and Utilities Alliance (EUA) paid a public relations company to run a smear campaign against electric heat pumps. Founded in 1905, the EUA is a trade association whose membership is heavily weighted toward the gas industry. Its mission is to "act on its members' behalf to shape the UK's decarbonisation policy whilst being mindful of affordability and security of supply." As we have seen, heat pumps are indeed an existential threat to gas heating. The campaign therefore aimed to "spark outrage" around heat pumps, and delay their roll out by confusing consumers with attention grabbing news headlines like "costly and noisy", "financially-irrational", and even "Soviet-style".

The campaign appears to have had some effect. A 2024 report by the UK's National Audit Office found that confusion around the role of hydrogen had contributed to the uptake of heat pumps being slower than desired. However, it may have done so at the cost of credibility, with the EUA reportedly receiving a ban from government meetings as a result of its propaganda.

In a March 2024 speech to Parliament, Energy Efficiency Minister Lord Callanan said: "I am supportive of a sensible debate on competing technologies, but planting misleading and false stories about heat pumps to negatively affect public support for the technologies is, frankly, a disgrace, and the big boiler manufacturers that fund the EUA should be ashamed of themselves."

The UKs gas network operators are also members of the EUA, and some have also been criticised. A key element of the government's evidence-gathering process was the so-called "hydrogen village" trials, in which small towns of 1,000–2,000 homes would be converted to 100% hydrogen for at least two years. Participants were offered cash incentives of £2,500–£3,000, new hydrogen appliances, and a fixed gas price during the trial.

However, there was no such price guarantee once the trial ended, nor any certainty that hydrogen supply would continue—unsurprising given hydrogen's expected cost premium and the lack of a government commitment to its long-term use in homes. Some experts speculated that gas network operators supported the trials in hopes of securing a legal basis for future investment, which could entitle them to compensation under UK law if the networks were later decommissioned.

Trials were planned for Redcar in northeast England and Ellesmere Port near Liverpool. However, residents in both towns rallied against these plans due to concerns around the safety of hydrogen in their homes, the future cost of hydrogen, and the feeling that they were being used as guinea pigs and that gas providers were not adequately answering their questions or addressing their concerns.

Kate Grannell, a resident and campaigner against the proposed trial at Ellesmere Port told *Hydrogen Insight* that they were subjected to "an unethical greenwashing campaign that has manipulated, misled and deceived our community." As a result of the public opposition, the government cancelled both trials – Ellesmere Port trial in July and Redcar in December of 2023. A third trial in Scotland is still on the table but has been delayed due to difficulties around securing the green hydrogen supply.

It is, of course, understandable that companies want to continue using infrastructure they have invested in. But that cannot come at the expense of the consumer and the climate. The UK case shows how corporate self-interest can obstruct public-good decision-making.

It appears that policy makers in gas-heating dependent nations are now drawing similar conclusions. Hydrogen heating was noticeable by its absence in the US 2023 clean hydrogen strategy and roadmap. Germany's 2023 Building Energy Act banned the installation of new gas boilers, but due to coalition-government politics, allowed for the installation of "hydrogen-capable" boilers under the condition that the local gas network operator guarantees that the supply will be switched to hydrogen by 2035. The German association of gas network operators has labelled this as "electrification through the back door" because a switch by 2035 is an impossible prospect, and one they will not even bother trying to meet.

Meanwhile the UK's National Infrastructure Commission recommended in 2023 that the government should not support hydrogen for home heating. At this time the UK government also clarified that heat pumps will be the primary low-carbon technology for decarbonising homes.

Taken together, the evidence is mounting: hydrogen is unlikely to play a substantial role in heating buildings.

Storing Sun and Wind

Renewable energy technologies are wonderful. They generate electricity – our most efficient and versatile energy carrier – with little to no greenhouse gas emissions. Wind and solar power in particular are now among the cheapest generation options available. They have one glaring limitation however.

Unlike fossil fuel and nuclear power plants, which can operate continuously, renewable energy generation is variable. Solar panels only work during the day, and windmills only generate when the wind is blowing. Neither of these conditions are constant, but our demand for electricity is. Also, electrical grids require electricity to be consumed at the same time it is produced, and the amount of electricity that wind and solar generate does not always align with demand.

One solution to these challenges is energy storage. Incorporating storage into electricity systems allows us to bank excess renewable energy when supply exceeds demand, and release it as required. Energy storage allows the decoupling of energy generation from consumption, providing a reliable and stable supply of clean power. It is a necessary component of a decarbonised power sector.

There are a few different timeframes, for which energy storage is required:

Short-Term Storage (Seconds to Hours): is required to address rapid fluctuations in demand or supply. These fluctuations can occur due to sudden changes in weather, or demand spikes.

Medium-Term Storage (Hours to Days): helps bridge the gap between renewable generation and demand, day and night, or short-lived periods of poor weather.

Long-Term Storage (Weeks to Months): is critical for managing seasonal variations in energy production. Typically, there is less sunlight in the winter months, and reduced wind during the summer. This type of storage is akin to "making hay while the sun shines" so that it is available during the dark days of winter.

Green hydrogen is often proposed as a medium for energy storage, but it is far from the only option. There are many competing energy storage technologies, each with their own strengths and limitations. Generally speaking, there is a trade-off between how expensive storage is to build, how efficiently it stores and releases energy, and how frequently it will be called upon.

Energy storage can be viewed as an alternative to electricity generation and must be priced similarly. Long term storage must be very cheap to build because it is not drawn upon frequently, making the number of uses per dollar invested low. In contrast, short term storage that is used for balancing supply and demand on a moment-to-moment basis will be called on many times per day. A more efficient but expensive technology may be acceptable here because the number of uses per dollar invested is high. Let's now look at the various technologies on offer, with an eye on where hydrogen may fit in.

Pumped Hydro Storage (PHS)

Pumped hydro storage (PHS) works by using surplus electricity to pump water uphill: from a lower reservoir to a higher one. When power is needed, the water is released to flow back down through turbines—just like in a conventional hydroelectric plant. PHS is a mature technology and by far the most widely deployed form of energy storage. In 2023, it accounted for two-thirds of global capacity by power rating.[14]

It's also among the most efficient, with roundtrip efficiencies of up to 85%. That means most of the input electricity is recovered, with only about 15% lost as heat. PHS is well suited to large-scale, long-duration applications and has been proven reliable for grid balancing over timescales of hours to days. Its main drawback is geography: it requires significant elevation differences between water bodies or costly dam construction.

Lithium-Ion Batteries (Li-ion)

Lithium-ion (Li-ion) batteries are everywhere—powering our phones, laptops, power tools, and electric vehicles. When connected to a circuit, they produce electricity by moving positively charged lithium ions from the anode to the cathode. Recharging reverses this process: an electric current pushes the ions back to the anode, ready for another cycle.

There are several Li-ion chemistries, but the type referenced here is lithium iron phosphate (LFP). While not as energy-dense as some alternatives, LFP is considered safer, longer-lasting, and cheaper – thanks to the absence of exotic metals. For these reasons, LFP batteries are often preferred for grid storage and are increasingly used in electric vehicles as well.

Li-ion is the second most widely deployed energy storage technology after pumped hydro, accounting for about 20% of utility-scale storage in 2023. Its fast response time makes it particularly well suited to short-duration applications.

The main drawbacks today are cost and competition from other sectors—particularly EVs. However, costs are expected to fall, and Li-ion use in grid storage is projected to grow rapidly. Unlike some other technologies, Li-ion systems are not geographically constrained, which further supports their adoption. The IEA expects Li-ion to be the fastest-growing energy storage technology in the coming years.

At present, the largest battery installations—located in California—can store around 3 GWh of energy, enough to power roughly 100,000 American homes for a day.

Flow Batteries

Flow batteries are an emerging technology with potential for medium- to long-duration energy storage. Unlike most other battery cells, which are built as self-contained units, flow batteries store energy externally in tanks of liquid electrolyte. When electricity is needed, the electrolyte is pumped through an electrochemical cell where the chemical reactions that generate electrical current occur. This provides design flexibility because unlike conventional batteries, the amount of power they can deliver and the amount of energy they can hold are not related.

The most common flow battery design uses vanadium (V) as the active element. Vanadium is particularly suited to this role because it can wear many different hats. With four different ionic forms—V^{2+}, V^{3+}, V^{4+}, and V^{5+}—vanadium can gain or lose between two and five electrons. In a vanadium flow battery, the electrolyte consists of the metal dissolved in an acidic solution. When electricity is required, the electrolyte is circulated through electrodes, triggering the following reactions:

Anode: $V^{2+} \rightarrow V^{3+} + e^-$

Cathode: $V^{5+} + e^- \rightarrow V^{4+}$

Here, e^- represents an electron and the reaction show how they are spontaneously and simultaneously created at the anode and consumed at the cathode. This push-pull relationship at the electrodes is what causes an electric current to flow.

The process can also be reversed by applying a current to the electrodes, driving electrons back the other way and 'charging' the tanks with V^{2+} and V^{5+}. It's a chemical analogue to pumping water uphill. In contrast to lithium-ion, flow batteries don't degrade with repeated use, making them well suited for applications requiring frequent cycling.

However, there are drawbacks. Vanadium is relatively scarce in the Earth's crust and is not produced in large volumes, which drives up cost and limits scalability. Flow batteries also have lower roundtrip efficiency than lithium-ion and a lower energy density, meaning they require more space. For now, they remain a fringe technology, although research into alternative chemistries is ongoing.

The largest operational flow battery to date is a 400 MWh vanadium system in Dalian, China, commissioned in 2022—about one-tenth the size of the largest lithium-ion installations.

Thermal Energy Storage

Thermal storage is an umbrella term for a variety of technologies that hold energy in the form of heat and release it either as heat or electricity. For grid-focused applications, the most relevant are the methods known as 'sensible heat' storage, where materials like molten salt, sand or oil are heated and stored at high temperatures. When electricity is needed, the stored heat is used to produce steam that drives a turbine generator. Alternatively, the heat can be delivered directly to buildings through a district heating network.

Thermal storage systems are typically best suited to large-scale facilities that produce substantial heat, such as concentrating solar power (CSP) plants, which use mirrors to focus sunlight onto a heat transfer material like molten salt. They are less useful for storing electricity from solar PV and wind, which generate electricity directly rather than heat. The efficiency of converting stored heat back into electricity is also relatively low compared to other storage options.

As a result, thermal storage will likely remain a niche technology, used primarily in compatible industrial settings or in regions with excellent solar resources. That said, the largest thermal storage facilities now have capacities comparable to the largest battery systems.

Gravitational Energy Storage

Gravitational storage is an emerging technology that is just taking its first steps on a commercial scale. Broadly speaking, it works by using excess electricity to lift heavy objects – like concrete blocks – to a higher elevation using a crane or winch. This requires either a vertical mine shaft or purpose-built tower. When electricity is needed, the blocks are allowed to drop under gravity and this movement drives an electric generator.

The concept is mechanically simple, environmentally benign, and well suited for medium to long-duration energy storage. Its main drawback is low energy density, as a lot of physical space is required for the mines or towers. Consequently, its viability is geographically constrained.

As with thermal storage, gravity storage seems likely to remain a niche technology.

Compressed Air Storage

Excess electricity can be used to compress air and store it in a sealed container. For small systems, this might be a tank or pipeline, but at utility scale, underground caverns are required. When electricity is needed, the pressurised air is released and used to drive a turbine—much like letting the air out of a balloon to spin a toy windmill.

One challenge with compressed air storage is heat loss. When air is compressed, the molecules collide more frequently, generating heat which is lost to the environment. This greatly reduces the available energy when the air is later released to drive a turbine. In simple terms, heat loss reduces roundtrip efficiency.

Early compressed air systems addressed this by reheating the air with fossil fuels before sending it through the turbine. This approach, while effective, undermines the goal of decarbonisation. A more recent innovation—known as adiabatic compressed air storage—captures the heat generated during compression using thermal storage technology. This heat is then used to warm the air during the discharge cycle. Adiabatic storage can provide a roundtrip efficiency of 70% – much better than the 40% typical of the older method.

The term 'adiabatic' comes from thermodynamics and means that no heat is exchanged between a system and its surroundings. In contrast, the older approach is known as 'diabatic' storage, where heat is lost or supplied externally.

The key advantage of diabatic compressed air storage is its relatively low cost. The main drawbacks are inefficiency and reliance on fossil heat. Diabatic systems have been in limited for decades. The world's first, the Huntorf plant in Germany, was commissioned in the 1970s and remains operational.

More recently, adiabatic systems have begun to come online, particularly in China. In May 2024, the world's largest compressed air facility at the time began operation in Shandong Province with 1.8 GWh of storage capacity. This is enough to supply electricity equivalent to a medium sized coal power plant for about six hours.

Hydrogen Storage

Hydrogen storage works in much the same way as compressed air. It is squirrelled away in suitable underground caverns and later withdrawn to generate electricity via fuel cells or combustion. The key difference is cost. Producing compressed air is cheap; producing green hydrogen is not. For medium-term storage, hydrogen is between one-and-a-half to three times more expensive than pumped hydro, and more than three times costlier than diabatic compressed air.

Worse still, hydrogen also has the lowest roundtrip efficiency of all the technologies discussed – and it's not close. After the energy losses from electrolysis, compression, and reconversion to electricity, only around 30% of the original renewable energy is recovered.

Given these drawbacks, it might seem mad to consider hydrogen for energy storage at all. Yet there are two compelling reasons why it may still play a role.

The first is scale. In a report with the self-explanatory title *Large Scale Energy Storage*, the UK's Royal Society concluded that hydrogen storage would be essential to achieving a zero-carbon electricity system by 2050. The report estimates that 60–100 TWh of storage capacity will be needed to support a renewables-dominant grid. This is roughly 1,000 times more than the UK's current pumped hydro storage capacity. The authors concluded that neither hydro nor batteries could feasibly scale to that level, but hydrogen stored in salt caverns could.

So why not use those same caverns for compressed air, given that it's a third of the price? The answer is energy density and this time, hydrogen is on the right side of that equation. At typical underground storage pressures of 100–200 atmospheres, compressed air holds 0.1–0.3 GJ/m^3, while hydrogen provides 1.5–3.5 GJ/m^3. That means hydrogen can store between 12 and 15 times more energy per unit volume than compressed air. For large-scale storage, that difference is decisive.

The second reason hydrogen may have a role is that its relative cost becomes more competitive for long-duration, seasonal storage. That case will be addressed shortly.

A small number of grid-connected hydrogen storage projects are now under development. The ACES Delta project in Utah is expected to begin operations in 2025 and will offer 300 GWh of dispatchable storage. This is roughly enough energy to power 10 million American homes for a day: around 8% of the population.

On a similar scale, the Grove Mulei hydrogen storage project in China's Xinjiang province began construction in late 2024. While exact storage figures are unavailable, the facility is designed to produce 110 tonnes of green hydrogen per day, which is about 10% more than ACES Delta. The stored hydrogen will be used for power generation and as fuel for heavy-duty trucks, which the developer also manufactures.

Power Storage Comparison

Table 6.4, presents data from the US Department of Energy's 2022 energy storage cost assessment. The levelized cost of storage (LCOS) measures the price a power storage facility would need to charge for a unit of electricity to cover all its costs, including taxes, financing, operations and maintenance. It's a way to compare the cost of owning and operating different sorts of power storage systems.

For comparison, the levelized cost of electricity from a natural gas-fired peaker plant in the US is $110–230 per MWh. A peaker plant generally only runs 10–20% of the time and is designed to ramp up power generation quickly, from a fuel source that is always available. Because it's not operating all the time, the unit cost of electricity from a peaker is substantially more than 'baseload' power, but the market accepts this as the price of continuous electricity supply. A natural gas peaker plant is the fossil equivalent of renewable energy storage, which as Table 6.3 shows, is often price competitive. Given that the US has some of the cheapest natural gas in the world, renewable energy storage may be even more competitive in other countries.

It is notable that the LCOS increases with the storage duration. This is because the stored energy is being used less frequently, but the investment required to build out the storage system is still the same, so we are getting less bang for our buck. Also note that the cost of technologies using naturally occurring storage like lakes and caverns are less sensitive to storage duration. This is because co-opting an existing geological feature is cheap compared to building storage media.

Ten hours of storage is a useful reference point because it corresponds to the lowest LCOS for all the technologies. It therefore lets us compare the technologies at

Table 6.4. Cost and efficiency of energy storage technologies at 1000 MW scale.

Storage Technology	Round Trip Efficiency	10 Hour LCOS ($/MWh)	100 Hour LCOS ($/MWh)
Compressed Air	52%	100	120
Pumped Hydro	80%	110	250
Gravitational	83%	120	430
Thermal	48%	150	370
Li-Ion Battery	83%	160	920
V Flow Battery	65%	180	1070
Hydrogen	31%	350	380

Source: US DOE (2022).[15]

their best. One hundred hours is more indicative of longer-term storage costs, where the return on investment is not as good.

Reflecting on the values presented we see that for a 10-hour duration, pumped hydro provides the best combination of efficiency and low cost. This is not surprising given that most existing storage uses this technology. Li-ion is 50% more expensive but slightly more efficient and its costs will come down. Compressed air is notable for having the lowest cost of all.

Hydrogen is dead last in round trip efficiency and in 10-hour storage cost. However, when it comes to 100-hour storage, hydrogen narrowly misses out on third place. Hydrogen storage is like an endurance runner; hopeless in track events, but does pretty well in the marathon.

Pumped hydro and compressed air are still more efficient and cheaper than hydrogen over both the 10-hour and 100-hour duration and should be preferred to hydrogen all else being equal. However, as we discussed they may not scale up sufficiently due to low energy density in the case of compressed air, or lack of suitable locations in the case of pumped hydro. Hydrogen could be the next best option.

Of course, this assumes that suitable underground storage is available for hydrogen, which it is not in all locations. Regardless, it is fair to conclude that hydrogen is at least worth considering for seasonal storage, but perhaps not for shorter durations.

Power from Ammonia

Most of the electricity we use today is produced using electromagnetic induction. Put simply, this is where a shifting magnetic field 'pushes' the electrons in a wire coil, causing electricity to flow. In a power plant, the dance between magnet and coil is brought about by a spinning shaft attached to a fan or rotor. The spinning action is created when something physically pushes against the blades of the rotor.

This could be high pressure air on a windmill or rushing water in a hydro plant. In thermal power plants, high pressure steam is used to spin the rotor and this steam can be generated by introducing water to any process that produces heat. Nuclear reactions, hot rocks in the earth, burning fossil fuels, and even the sun are all used commercially to drive steam turbines.

Gaseous fuels can even be burned directly in a gas turbine, which is a close relative to the jet engines used in aeroplanes, sidestepping the need for steam entirely.

In theory, ammonia is a fuel that could be burned directly in a gas turbine, or in a boiler to produce steam. Doing so would be of interest to countries wanting to import clean energy because, as we have seen, ammonia is a more feasible energy carrier than hydrogen. However, ammonia's poor combustion characteristics mean that this is easier said than done.

Despite research dating back to the late 1960s, commercially available gas turbines and boilers that can run on ammonia do not currently exist. However, there are a handful of companies currently trialling megawatt scale turbines that can run on pure ammonia or ammonia/methane blends, with hopes of making them available in the near future.[16]

The second obstacle to ammonia power is that the green molecule is still very expensive. The cost in 2025 sits in the approximate range of $800–1000 per tonne. Modelling indicates that if the cost of green ammonia can be reduced by a half or more, using it in a gas turbine to produce electricity could be an economical way to balance a power grid.[17] At a price of $380/t NH_3, the levelized cost of electricity from an ammonia turbine was estimated at $150–$200/MWh. This is similar to the price of electricity from the natural gas peaker plants used today.

So if the combustion technology can be perfected, and the price of green ammonia can be greatly reduced, it is feasible that ammonia turbines could play a role in future green grids – providing power during peak times, or when the sun isn't shining and the wind isn't blowing.

In parallel to the development of ammonia turbines, the idea of co-firing ammonia with coal in existing power plants is receiving interest throughout coal-reliant countries in Asia. This idea is particularly big in Japan, which is heavily involved in its development and promotion.

The world's first industrial scale demonstration project took place during 2024 at the Hekinan power plant near Nagoya. Jera, the company behind the trial, hopes to have the co-firing technology commercially available by the middle of 2025. Meanwhile, the Japanese government is laying the path for its companies to export the technology. Japan's *Asia Energy Transition Initiative* offers $10 billion in financial support for decarbonisation technology, with prominent emphasis on ammonia co-firing.

In addition to Japan, co-firing projects have been announced in Bangladesh, China, India, Indonesia, Malaysia, the Philippines, Singapore, South Korea, Taiwan, Thailand, and Vietnam. The Chinese government has also decreed that its coal-fired power plants must begin to reduce their greenhouse emissions in 2025, with the goal of 50% reduction by 2027. Co-firing green ammonia is one of the options they have been given along with co-firing biomass, and carbon capture and storage.

Unsurprisingly, the very idea of co-firing ammonia and coal has been proven controversial. The narrative of the Japanese Government and the companies promoting the technology is that, nations reliant on coal for their power can simply apply a few tweaks and continue to use their existing power stations. Starting with a blend of 20% clean ammonia and 80% coal, generators can gradually increase the proportion of ammonia over time (although this has not yet been demonstrated) before finally arriving at 100% ammonia sometime between 2040 and 2050; just in time to meet everybody's net-zero pledges. It is argued that this is a way to quickly reduce emissions from fossil-fuelled power generation in the short term, but there are several criticisms of this plan.

The first is that emissions reductions are modest, as Table 6.5 shows. Burning diluted coal is still burning coal, and decarbonisation implies not doing this. A plan that locks nations into coal-power for decades to come is incompatible with the goal of transitioning to net-zero emissions by 2050.

Secondly, clean ammonia is currently not widely available, so the assumption that, sufficient volumes can be secured in the near term is highly suspect. The Hekinan demonstration project actually used grey ammonia, which provides no

Table 6.5. Cost of decarbonising power generation with ammonia.

Fuel Mix	CO_2 emissions (t/MWh)	2024 LCOE ($/MWh)	2030 Projected LCOE ($/MWh)	2024 Abatement Cost ($/t CO_2)	2030 Abatement Cost ($/t CO_2)
100% Coal	0.9	80	80	--	--
Green Ammonia					
20%	0.72	110	85	170	30
50%	0.45	190	135	240	120
100%	0	320	225	270	160
Blue Ammonia					
20%	0.74	100	90	125	65
50%	0.50	160	155	200	190
100%	0.10	270	265	240	230

Source: BloombergNEF (2022).[18]

climate benefit because the emissions from making the ammonia are about the same, if not more, than the emissions from burning coal. Energy analysts at the firm BNEF have calculated that if all the projects announced by the 12 countries listed previously were to co-fire with 20% ammonia, they would consume 24.4 tonnes per year. That is 35% more than the 18 million tonnes of ammonia that were shipped worldwide in 2023. Building up the production and shipping capacity to allow for export of clean ammonia on this scale would take many years and perhaps decades. This appears to neuter the argument that co-firing is something that can be implemented quickly to get emissions down in the short term.

The third major criticism is that clean ammonia is very expensive compared to coal. The ammonia price mentioned earlier translates to $150–200/MWh, while coal suitable for power generation costs around $15–30/MWh. Obviously, adding ammonia to the mix will increase the total cost of fuel that a power plant operator must pay. There is more to the price of electricity than the fuel cost however. Power plants must also be financed, built and maintained, and in the case of co-firing, modified.

Putting all this together, BNEF has estimated the levelized cost of co-fired electricity in Japan, and these figures are provided in Table 6.5. They assume that green ammonia is shipped to Japan from Australia and that blue hydrogen is shipped from the Middle East. Keep in mind that these can only be estimated as neither commercial co-firing nor clean ammonia supply chains currently exist outside of trials.

Table 6.5 also shows the results of back of the envelope calculations for CO_2 abatement costs. Here we assume that green ammonia produces no emissions, and that blue ammonia has a footprint of 0.1 tCO_2/MWh – the value that was quoted for the world's first shipment of certified blue ammonia, which took place in May 2024 between the UAE and Japan.

With 2024 pricing, a 20% ammonia blend is expected to raise the cost of electricity by at least 25%. If we treat that premium as the price of emissions reductions, it translates to an abatement cost of around $170 per tonne of CO_2 for green ammonia, and $125 per tonne for blue.

These figures fall within the realm of economic plausibility in places with high carbon prices. Uruguay, for example, has the world's highest carbon tax at $167 per tonne, while several European countries fall between $100 and $130. In these locations, 20% ammonia co-firing might be justifiable. But in most of the world, carbon pricing remains far lower – Japan's is just $1.90 per tonne.

At higher blend ratios, the economics quickly unravel. By the time we reach 100% ammonia, the levelized cost of electricity becomes exorbitant, and the cost of avoided emissions surpasses $200 per tonne.

Looking ahead to 2030, the picture improves slightly—assuming that the price of green ammonia drops significantly while coal power remains stable. Under those assumptions, a 20% green ammonia blend could achieve near parity with coal, making abatement very affordable. Blue ammonia might follow a similar trend, though the gains would be smaller because the technology is mature, and costs are unlikely to fall much further. In either case, pushing beyond a 20% blend remains costly. Even if the carbon abatement looks tolerable on paper, the associated electricity price likely won't, unless coal is subject to a much more punitive carbon price than exists today.

Some energy analysts argue that renewables, which offer meaningful emissions reductions, are less costly options than co-firing ammonia. BNEF calculates that the levelized cost of electricity from onshore wind and solar generation plus expensive battery storage, costs about the same as 50% ammonia co-firing now, and will be much cheaper in the future as the cost of both renewables and storage continue to come down.[18] Analysts from non-profit think tank Transition Zero make much the same argument. They calculate the cost of solar and onshore wind energy deployed in the Philippines, Malaysia, Indonesia and Thailand to be in the range of $50–$90 per tonne of CO_2 avoided. This is 50–70% less than the cost of emissions reductions from co-firing.

Considering the low availability of clean ammonia, the modest emissions reductions and the high price of co-firing it is easy to see why accusations of greenwashing have been levelled at the parties promoting it. Arguments have been made that they are in fact motivated by the desire to continue profiting from existing coal power stations, and from selling the co-firing technology, rather than a genuine desire to decarbonise.

Friction around this issue has also been noted at the ministerial level. At the April 2023 G7 meeting of Climate, Energy, and Environment ministers held in Japan, the hosts resisted calls to phase out coal by 2035 but were keen to have ammonia co-firing recognised as "an effective emissions reduction tool". The Financial Times reported that Canada unsuccessfully tried to replace the word "effective" with "potential", while the UK and France were able to water down the language by qualifying that it "should only be used if it aligns with climate targets". The final passage from the official communique hints at these diplomatic wranglings:

> "We recognize low-carbon and renewable hydrogen and its derivatives such as ammonia should be developed and used where they are impactful as effective emission reduction tools to advance decarbonization across sectors and industries, notably in hard-to-abate sectors in industry and transportation. We also note that some countries are exploring the use of low-carbon and renewable hydrogen and its derivatives in the power sector to work towards zero-emission thermal power generation if this can be aligned with a 1.5°C pathway and our collective goal for a fully or predominantly decarbonized power sector by 2035, while avoiding N_2O as a GHG and NOx in general as a regional air pollutant and precursor to tropospheric ozone."[19]

While we can't say how aggressively the co-firing strategy will be pursued in the coming years, it is quite clear that not everybody is onboard and that it's very unlikely to contribute to a fully or even a predominantly decarbonised power sector by 2035.

Big Numbers: Chapter 6

- **40%** - Share of global building heat provided by natural gas: a major decarbonization opportunity.
- **4x** - Predicted increase in home explosions if hydrogen replaces natural gas in residential supply.
- **5x** - Additional renewable electricity required for green hydrogen heating relative to electric heat pumps.
- **30%** - Round trip efficiency of hydrogen power storage: the lowest of available options.
- **100 hours** - Timescale at which hydrogen power storage becomes comparatively affordable: a candidate for high-capacity, long-duration storage.
- **50%** - Price reduction needed before green ammonia becomes viable for grid balancing. Combustion technology must also be commercialised.
- **3.5x** - Cost of electricity from ammonia compared to coal: co-firing is an expensive strategy.

CHAPTER 7

Hydrogen in Transport

∞∞∞

Hydrogen on the Move

Transport relies on oil products for over 90% of its energy, which contributes around 8 giga-tonnes (Gt) of carbon dioxide to the atmosphere every year. This is roughly 14% of global greenhouse gas emissions, and our demand for transport is growing all the time. The IEA states that emissions increased by nearly 2% year on year between 1990 and 2022. There can be no argument that decarbonisation of the global economy requires a massive shift to low-carbon transportation systems.

This can be partially achieved through emphasising the most efficient transport modes. As Figure 7.1 illustrates, the use of trains, buses, and small personal vehicles

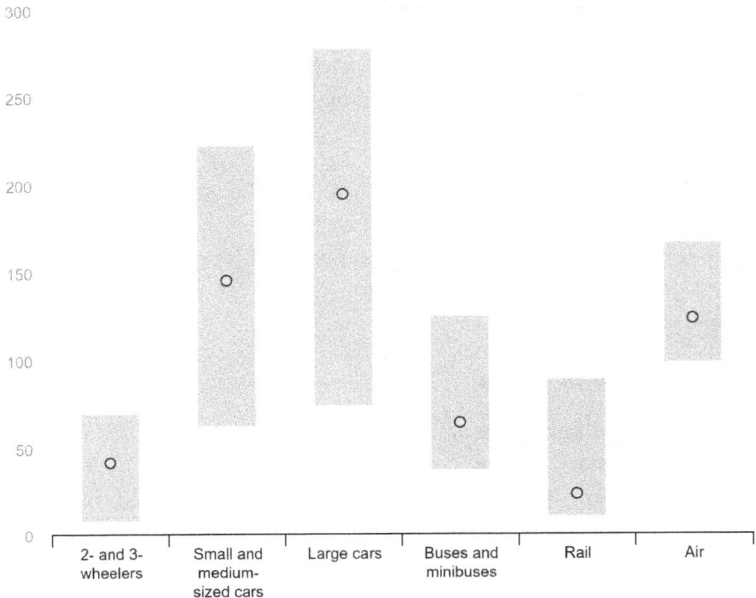

Figure 7.1. Well to wheel greenhouse gas intensity of passenger transport modes. Values are in g CO_2-eq/ passenger-km. The bars show the global range, the dots show the global average. Source: IEA (2022).[1]

DOI: 10.1201/9781003361428-9

like scooters, results in substantially fewer emissions than planes and cars. When it comes to freight, shipping and rail require just 10% of the energy of trucking per tonne-kilometre.[2]

Prioritising public transport, along with sea and rail, is therefore an obvious way to address energy usage and emissions. To achieve deep emissions cuts however, in addition to using the most energy efficient means of transport, we must decarbonise the energy that powers them. Hydrogen has long been touted as a means of achieving this and it can be applied in a few different ways.

Hydrogen Fuel Cells

The first, and most attention-grabbing method is the use of fuel cells. A fuel cell is conceptually the reverse of an electrolyzer. It combines high-purity hydrogen from storage with oxygen from the air to produce a flow of electricity, with water as the only byproduct. In vehicles, this electricity powers an electric motor to drive propulsion. Hydrogen fuel cell electric vehicles (FCEV) are considered zero-emission because nothing comes out of the tailpipe other than a dribble of water or plume of steam, depending on the temperature outdoors.

A pedant might point out that water vapour technically is a greenhouse gas, so FCEVs are not strictly zero emission. However, water vapour is not generally considered a driver of climate change given that it is part of the natural hydrological cycle. Water lasts only a few days in the atmosphere before raining back down to earth, in contrast to greenhouse gases of concern like carbon dioxide, methane, and nitrous oxide that survive in the atmosphere for decades or centuries.

In addition to climate benefits, fuel cells also use fuel more efficiently than internal combustion engines (ICE). Under optimal conditions, stationary proton exchange membrane (PEM) fuel cells can transform around 50–60% of the energy contained in hydrogen into electricity. Stationary fuel cells are large and used to supply the grid or provide backup power. Mobile fuel cells suitable for vehicles are less efficient at 40–50% – still better than ICEs, which are heavily taxed by nature. Thermodynamic limits cap petrol engine efficiency at around 30%, with diesel engines closer to 50%.

This combination of energy efficiency and zero emissions are often highlighted by those wishing to establish the environmental credentials of fuel cell electric vehicles.

Hydrogen Combustion Engines

Hydrogen can also be burned in an internal combustion engine using technology much the same as that used in conventional vehicles today. A hydrogen ICE (HICE) is not really a zero-emission vehicle however, as burning hydrogen in air produces nitrous oxides (NOx), which are a major contributor to air pollution. In fact, a hydrogen engine has the potential to produce more NOx than a fossil fuel engine because hydrogen burns hotter. This is something that must be addressed by engineering design and engine maintenance.

As we've just noted, ICEs are less efficient than fuel cells. The best HICEs can match the peak efficiency of a good diesel engine, but under real-world conditions they are generally more wasteful than FCEVs. All else being equal, fuel cells remain the more efficient and environmentally preferable way to use hydrogen.

Derivative Fuels

As described in Chapter 4, clean hydrogen can be used to produce low-carbon versions of hydrocarbon fuels that closely resemble their fossil counterparts. These hydrogen-derived fuels include green ammonia, green methanol, renewable diesel, biojet fuel, and electrofuels such as synthetic diesel, gasoline, and kerosene.

Some of these can be used in fuel cells, and most can be burned in existing internal combustion engines—a convenient feature for those who already own such vehicles.

Each fuel has a unique set of properties and challenges, and each transport mode has its own requirements. Let's now look at which combinations are most promising for land, sea, and air.

Automotive

If you were to stop a random person in the street and ask them what they know of hydrogen, chances are they would suggest that it is a futuristic fuel for cars. This likely stems from FCEVs being the main focus during earlier waves of hydrogen enthusiasm in the 1980s and 2000s. In the past, interest in hydrogen-fuelled transport was driven by concerns around energy security and price. While these concerns still hold, the main motivation today is of course climate. Greenhouse gas emissions from road transport in 2022 amounted to nearly 6 Gt, which was 75% of transport emissions. Consequently, a switch to zero emission vehicles (ZEV) represents a golden decarbonization opportunity.

FCEVs have been shown to offer good to great emission reductions compared to ICE vehicles. Of course, this depends on how the hydrogen is made. To determine how climate friendly a vehicle truly is, we must include the emissions from producing and distributing the fuel in addition to the emissions from the vehicle itself. We must apply well to wheel analysis (WTW).

A recent US study[3] found that the Toyota Mirai FCEV emits about 28 g CO_2/km when fuelled with green hydrogen, compared to 228 g CO_2/km for a similarly sized Mazda 3 running on petrol. If the hydrogen comes from natural gas instead, emissions rise to 144 g CO_2/km. That's a WTW saving of nearly 90% for green hydrogen, but only around 30% for grey.

A separate but complementary South Korean study[4] has shown that electrolytic hydrogen must be produced using clean electricity if FCEVs are to provide carbon savings. Korea's electricity grid is dominated by fossil fuels, particularly coal, and the researchers found that using it to produce hydrogen resulted in FCEV emissions of 380 g CO_2/km. This is 70% more than an equivalent petrol car. Similar results were reported for Europe and the US, where grid electricity would lead to FCEVs emitting

340 g and 517 g CO_2/km, respectively. Electrolysis is highly energy intensive, so when that electricity is dirty, the resulting hydrogen is too.

These combined studies make one thing clear: without clean hydrogen, FCEVs offer little to no climate benefits.

So, assuming our hydrogen is green, are FCEVs our best hope for eliminating those 6 Gt/yr of CO_2 emissions? As it turns out, no. There is another.

Electric vehicles that run off a battery (BEV) also offer a zero-emission drive train, with the added advantage of superior efficiency. This is demonstrated by Table 7.1, which provides a WTW comparison using representative values for each link in the fuel supply chain. We can see that BEVs benefit from an exceedingly simple conversion process, which greatly helps efficiency. Electricity generated from renewables is transmitted almost instantly as alternating current (AC) through power lines to the car charger where it's transformed to direct current (DC) and loaded into the battery. This energy is then available to drive the wheels via a highly efficient electric motor. That is only four steps. Four gates at which nature's tax collector can clip the ticket. What's more, each step is efficient, meaning the 'tax' rates are relatively low.

By comparison, the journey from well to wheel is more convoluted for FCEVs. Hydrogen must first be produced via electrolysis, then compressed for transport, trucked to the service station, and compressed again to fill the vehicle before finally being converted back to electricity in the fuel cell. In addition to having more steps, it includes two that are particularly inefficient: electrolysis and the fuel cell itself.

Technological improvements might shave a few percentage points off individual steps, but they are all bound by physics and the overall result will not change: approximately 70% of the original renewable electricity is lost to the ether with a FCEV, compared to less than 30% with a BEV. Put another way, the quantity of renewable generation we would need to power a fleet of FCEVs would be two and a half to three times greater than that required by an equivalent fleet of BEVs. That's a choice between building 100 wind turbines, or 250–300 wind turbines for the same amount of mobility.

Table 7.1. Well to wheel efficiency comparison of battery electric and hydrogen fuel cell vehicles.

FCEV	Step Efficiency	BEV	Step Efficiency
1. AC-DC Conversion	95%	1. Grid transmission	90%
2. Electrolysis	75%	2. AC-DC Conversion	95%
3. Compression	90%	3. Battery charging	95%
4. Transport (truck)*	98%	4. Electric motor	90%
5. Fuel Cell	50%		
6. Electric motor	90%		
Total Efficiency	28%	Total Efficiency	73%

Source: Bossel (2006).[5] *Assumes FCEV truck with compressed H_2 on a 200 km round trip with fuel efficiency of 10 km/kg H_2.

Maybe we have the space and means to build 300 turbines anyway. In that case, do we choose to use them all for cars, or do we use 100 of them for cars and 200 for powering our homes and businesses with clean electricity, replacing fossil-fired power generation? If the goal is to decarbonise as quickly as possible, the choice is obvious.

Of course, as with FCEVs, how much BEVs contribute to decarbonisation depends on where their electricity comes from. According to the IEA, WTW emissions from a BEV can range from zero in countries like Iceland, where electricity is nearly 100% renewable, to 187 g CO_2/km in South Africa, where coal dominates the grid.[6] A more nuanced comparison is provided by Figure 7.2, which shows WTW emissions for three different sizes of car fuelled by petrol, diesel, and electricity. For a global average grid, BEVs provide emission savings of 45% compared to petrol cars. This drops to the 20–25% range for a dirty grid, represented here by Australia. That said, a BEV owner who also happen to live in one of the roughly 40% of Australian homes with rooftop solar, could achieve near-zero WTW emissions – similar to the Norwegian average.

The only real advantages that FCEVs have over BEVs has been fuel capacity and the time it takes to refill. Hydrogen's volumetric energy density, while low compared to fossil fuels, is still better than current batteries. When compressed to 700 atmospheres, which is common in hydrogen fuel tanks today, hydrogen can hold

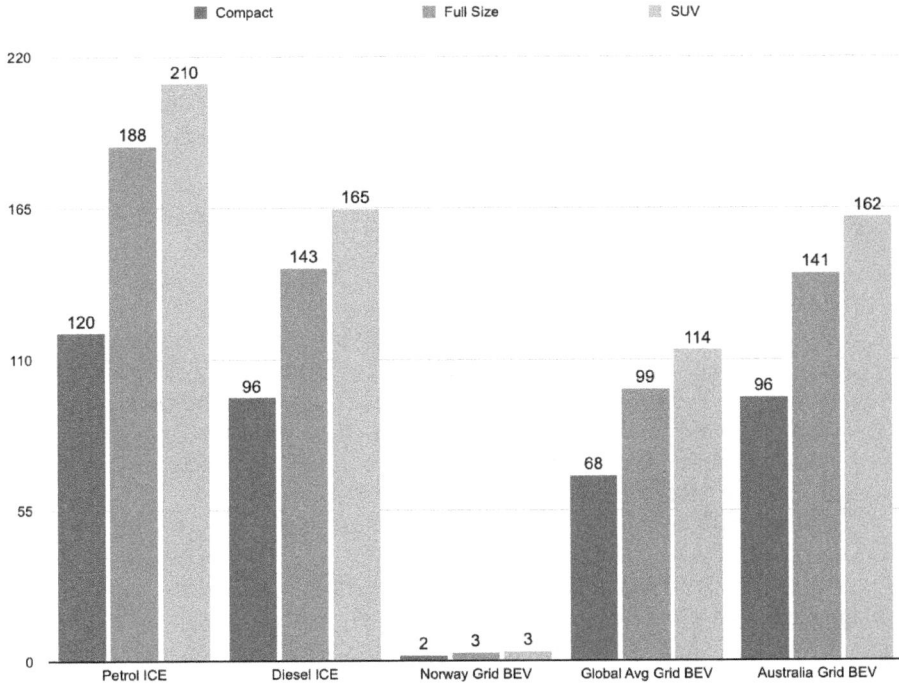

Figure 7.2. WTW emissions of ICE cars and BEVs considering electricity generation mix (gCO_2/km).
Source: Woo et al. (2017).[7]

1550 watt-hours per litre (Wh/L). In contrast, an EV battery will have a density closer to 250 Wh/L. However, half of the energy in the hydrogen will be lost in the fuel cell, so a litre of compressed hydrogen will only provide around 775 Wh. Hydrogen may hold six times the energy of a battery in a given storage space, but a FCEV's effective advantage is closer to three times.

Furthermore, storage space is not equal. Onboard hydrogen storage tanks are limited by engineering considerations and safety requirements. They require expensive materials to safely contain the high-pressure gas and increasing capacity adds weight and cost faster than it adds range – effectively capping tank size. BEV storage is different: limited only by the physical space available for battery packs and the price a consumer is willing to pay for them. As the price of batteries has come down, the range that automakers are able to provide has increased.

When it comes to refilling (or charging) time, physics favours hydrogen. Like inflating a tyre or pumping gas, refilling a hydrogen tank is simply a transfer of pressurised fluid – albeit at cryogenic temperatures if stored as a liquid. Charging a battery is slower, but not because electrons are slouches – in fact, it's the opposite. Fast flowing electrons generate heat as they jostle on an atomic scale. Inside a battery, this heat can trigger unwanted chemical reactions that degrade the materials and shorten its lifespan. To protect the battery, charging systems must manage the temperature by restricting flow, and this slows the process down.

The degree to which these advantages move the needle towards FCEVs depends on the application. Private cars, which spend 95% of their life stationery clearly have different requirements to working vehicles like a public commuter bus or a 600 tonne mining truck. Given the potential for deep emissions cuts and the fundamental efficiency advantage of BEVs over FCEVs, the question isn't which technology we should use to replace ICE vehicles—it's whether BEVs can do the job. Let's examine this now.

Personal Mobility

During the first half of the 2010s – the early days of mass market electric vehicles – FCEVs provided 400–500 km of range, while BEVs would struggle to give you 135 km. Range was clearly a heavy advantage for FCEVs at that time. However, thanks to reductions in battery costs, the BEV offering has markedly improved and is now reaching parity. In 2025, the newest Hyundai Nexo and Toyota Mirai claim ranges of 700 km and 650 km respectively, while most new BEVs have a range between 400 and 500 km, with premium models offering 600–700 km. The range advantage for hydrogen cars has all but dried up.

Similarly, refilling time still favours hydrogen, but the gap is closing. Both the Nexo and Mirai take around 5 minutes to fill up, which is 125 km of range per minute of refuelling time. The rate of battery charging depends on a particular car's battery management system and the power throughput of the charger, which varies greatly. Currently available fast chargers top out at around 350 kW, but only a limited number of high-end models are capable of drawing this. The most common DC fast charger provides 50 kW, while 150 kW is becoming increasingly common. Thus, most BEV drivers topping up during a long journey can expect to gain between 5 and 15 km of range per minute of charging, assuming a typical BEV efficiency of 6.5 km/kWh. So, hydrogen cars still refuel at least eight times faster than most battery cars can charge.

Whether this advantage is decisive depends on the needs of the owner, but for most, the current generation of BEV is quite adequate. Most of the time, these vehicles will be charged at the owner's home overnight while they are sleeping. In this case, it matters little if charging takes many hours. Also, private cars average around 60 km per day, which is well within the range of any BEV – even the very first models. Charging speed only becomes relevant when a motorist is on a long-distance journey.

Assuming a BEV has a range of 350 km, a 150 kW fast charger can give the battery an 80% top-up in about 15 minutes, which is not much slower than filling a FCEV. This would give the car another 280 km of range: another two-plus hours of continuous driving under a speed limit of 100–120 km/hr. This is a convenient distance because it is widely recommended that motorists on long journeys should stop for a rest every two hours to reduce the risk of fatigue related accidents. So, for most individuals the currently available charging technology - also rapidly improving – is clearly fit for purpose.

Access to refuelling infrastructure is also easier for BEV owners. In most locations, they can be charged (slowly) using existing domestic power outlets. Notable exceptions are North America and Japan, where the public electricity supply voltage is around half that of other countries, making charging from a conventional socket impractically slow. Regardless, dedicated home chargers can be installed for a small fraction of the car price. With the caveat that this is not always practical in apartments, BEVs simply do not require special infrastructure in the same way as FCEVs. What's more, chargers are cheaper and easier to install than hydrogen filling stations. As of 2023, there were just over 1100 filling stations globally, compared to 1.4 million public fast chargers.

In 2023, buyers had nearly 590 models of BEV to select from and elected to purchase nearly 10 million of them: 12.5% of all new car sales.[8] Energy consultants BNEF have reported that global sales of new ICE vehicles peaked in 2017 and any growth in overall automobile sales is now due to increasing rates of electric vehicle purchase. When it comes to personal mobility, consumers agree that BEV can do the job. The global fleet is electrifying.

Commercial Vehicles

The combination of range and faster refiling naturally leads to the suggestion that hydrogen may be preferable to batteries where profitability depends on keeping vehicles moving. Hydrogen may also be more suitable for vehicles that require a lot of power to deal with heavy loads. Power is of course the word we use to describe the amount of energy used per unit of time. Operators of trucks and buses may accept hydrogen's higher cost if it enables more hours on the road.

While it is still early days for zero-emission heavy transport, the relative advantage of fuel cells is similar to that of cars. For example, before it went out of business, the start-up automaker Nikola offered both hydrogen and battery electric trucks. Their hydrogen model claimed an 800 km range with a 20-minute refill time, while the battery version provided 500 km and a 90 minute recharge at 350 kW. The hydrogen version therefore offered 60% more range and significantly less downtime.

However, the hydrogen premium is not small. Based on findings from a UK-based heavy-freight demonstration project, engineering Professor David Cebon from the University of Cambridge has stated that a 40 tonne hydrogen truck is around twice the price of a battery equivalent, and would cost 2.5–3 times more to run. It is expected this will more or less always be the case due to the price differential between green hydrogen and electricity, and because the fuel cell drive train is more complex and includes costly elements like storage tanks, hydrogen delivery systems, and the fuel cell itself.

The situation is similar for buses, as shown in a study from the Italian city of Bolzano.[9] Researchers tracked real-world energy use of battery electric and hydrogen fuel cell buses over 15 months and found that the hydrogen models consumed 3.1–3.4 kWh per kilometre, while battery versions used just 1.4–1.5 kWh/km. Based on local fuel prices, this meant operating costs of $0.60/km for the battery buses and $1.37/km for hydrogen—more than twice as much, even before factoring in the higher maintenance needs of fuel cell systems. Numerous municipalities around the world have cancelled or rolled back plans for hydrogen bus fleets after coming to a similar conclusion.

The cost of hydrogen vehicles appears to be taking its toll on manufacturers. In early 2025, French bus maker Safra was placed into receivership after shifting focus to hydrogen, ending 70 years in business. Belgian firm Van Hool had gone bankrupt just 18 months earlier. The year 2024 was particularity dire, seeing the collapse of five hydrogen truck-makers—Nikola and Hyzon in the US, Tevva and Arrival in the UK, and Germany's Quatron. As energy economist Wolf-Peter Schill of the German Institute for Economic Research observes, battery-electric trucks in Germany now outnumber hydrogen models by 400 to 1—a stark illustration of the market reality these companies faced.

It's worth noting that the German study focused on regional trucking with ranges of 200–300 km. Appetite for hydrogen may simply vary by geography. For instance, Cambridge University's Centre for Sustainable Road Freight found that all current UK haulage could be handled by BEVs with sufficient charging infrastructure. In contrast, the North American Council for Freight Efficiency expects hydrogen to play a role in Canada and the US, where routes are longer. They estimate battery trucks can manage trips up to 400 km with 20-tonne loads; beyond that, hydrogen might be needed. This view is echoed by Norwegian consultancy DNV, whose 2050 forecast sees just 1% of global road transport powered by hydrogen, with 60% of it in North America.[10]

While start-up hydrogen truck makers have been struggling, many of the legacy manufacturers are eyeing up another option – hydrogen internal combustion engines. These have not been considered for cars in any serious way, but they are drawing interest from major truck makers like Volvo and MAN, who are concerned with maintaining existing production lines as demand for traditional ICE vehicles winds down. It is believed that some customers will prefer HICE trucks because they are half the price of a fuel cell truck and are a familiar technology. Even though a HICE is less fuel efficient, it takes around half the working life of the truck for the additional operating cost to outweigh the initial savings. By this time, the cost of better technologies may have come down and the company will have had more experience handling hydrogen. HICE can therefore be viewed as a way for risk-averse companies to dip their toes into decarbonisation. India is particularly active, with two of its truck-makers pursuing HICE. Tata Motors has a partnership with engine-maker Cummins and began production in 2024, while truck builder Ashok-Leyland has announced plans to do the same by late 2026.

Another sector where hydrogen is often proposed as a solution is off-road vehicles – the rugged workhorses of construction, mining, and agriculture. It is often assumed that batteries cannot provide the power needed to operate these monsters. It is perhaps surprising then, that three of the world's largest mining companies – BHP, Rio Tinto, and Fortescue – have all said that their preference is battery electric for haul trucks, as reported by *Hydrogen Insight*. Mining haul trucks are currently the biggest user of diesel in Australia.

Reasons given included the superior energy efficiency of batteries, and avoiding unnecessary building and land disturbance associated with the extra power generation and transmission that hydrogen would require. While battery electric mining trucks are still under development and not yet commercially available, BHP has stated that in their view "an electrified mining fleet is more economic and more achievable than the alternative fuel sources".

Uptake

In 2023, almost 94% of the world's FCEVs were located in just five territories: California, China, Korea, Japan, and Germany. Most of the remaining 6% are also to be found in Europe, with only a few hundred scattered throughout the rest of the world. California is the only state in America with a meaningful stock of FCEVs and fuelling infrastructure, which is why it makes the list.

Figure 7.3 shows how these vehicles are distributed. We can see that South Korea was the leader in wheels on the road. This is down to the substantial effort made by Hyundai—whose Nexo was the only FCEV available—and to generous government subsidies, which slashed the purchase price of a new Nexo by nearly half.

Rapidly closing in is China, who overtook Korea in annual FCEV sales for the first time in 2023. China is an outlier due to the composition of its FCEV fleet. While all other territories are car dominant, China hadn't cracked 1000, yet it ranked second overall thanks to 10,000 trucks, 6500 buses and nearly 3000 vans. This is because China's fleet is largely the result of strategic government policy directives and subsidies, rather than natural market development. China takes the position that it should be involved in all emerging technologies, and that includes hydrogen vehicles.

However, the sales of hydrogen vehicles need to be placed in the context of total zero-emission vehicle uptake. By 2023, around 88,000 hydrogen fuel cell vehicles were on the road worldwide—a figure dwarfed by the 28 million battery electric vehicles that now quietly dominate the zero-emission landscape. Table 7.2 presents the number of registered BEVs and FCEVs in each of the five territories where fuel cell vehicles have a presence. The numbers speak for themselves. Even in South Korea – the country that has most embraced FCEVs – batteries are preferred at a rate of 14 to 1. Japan has the lowest ratio, with 'only' four BEVs to every FCEV, but it also has the lowest number of ZEV by a long way. At 21,000 to 1, China's ratio is greater than the odds of being struck by lightning.[1]

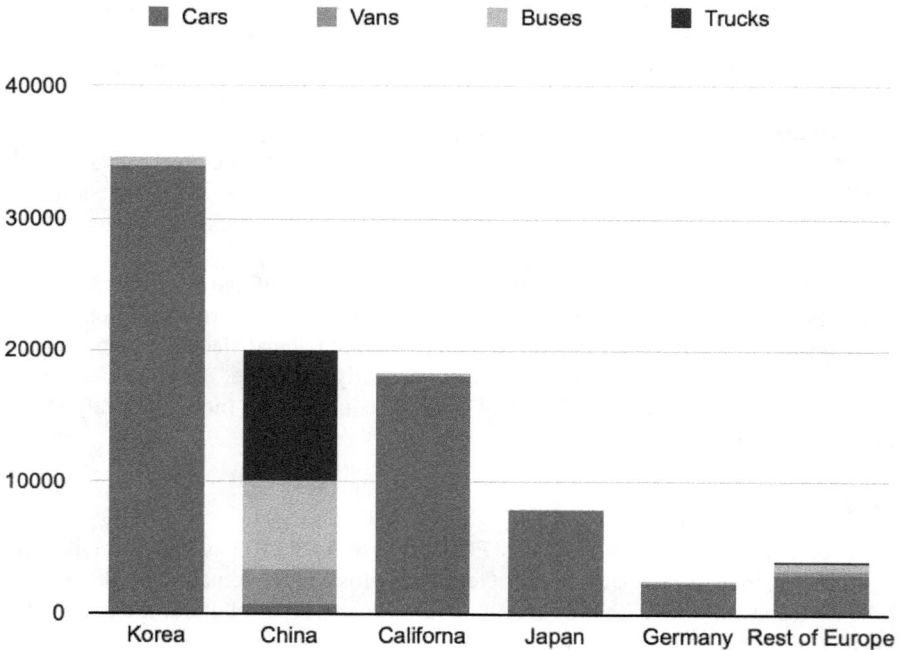

Figure 7.3. Global Distribution of FCEV in 2023. Source: IEA (2024).[11]

Table 7.2. Total battery electric and hydrogen fuel cell cars in major markets as of 2023.

Territory	Ratio of BEV:FCEV	Total Hydrogen Fuel Cell Cars	Total Battery Electric Cars
Japan	4	7900	290,000
South Korea	14	34,000	460,000
California	76	18,000	1,300,000
Germany	625	2400	1,500,000
China	21,000	800	16,000,000

Source: IEA (2024).[10]

---

[1] According to the US National Weather Service.

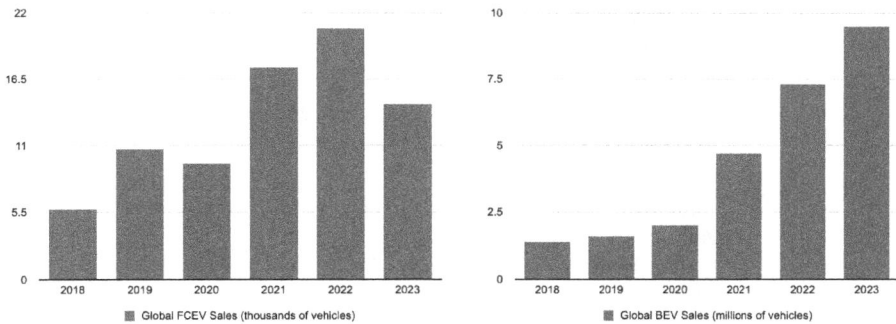

Figure 7.4. Global sales of fuel cell and battery electric vehicles: 2018–2023. Sources: Statista (2024), IEA (2024).[10]

Figure 7.4 shows global sales of each drive train for the period 2018 to 2023. It's important to note that they cannot be placed on the same graph because FCEV sales would be invisible due to the difference in scale – like trying to look at an acorn on the ground and the tree it fell from simultaneously. A quick glance shows that BEVs are going from strength to strength, with rapid growth in annual sales. The IEA anticipates that 40–65% of new vehicle sales in 2030 will be BEV. On the other hand, FCEV sales have been erratic—plunging in 2023 and reportedly falling a further 20% in 2024, which is not included in Figure 7.4.

The collapse in FCEV sales is likely due in part to the growing number and quality of BEVs on the market, while the FCEV offering outside China remains limited to just two models: the Toyota Mirai and Hyundai Nexo. Confidence has also taken a hit in key markets such as Korea and California, where recent hydrogen supply disruptions left some stations offline and others with hour-long queues. These shortages also triggered price spikes, exacerbating the already high running costs relative to other options.

Compounding these challenges, hydrogen refuelling stations have been shutting down in several markets due to poor economics. Shell closed the stations in California and the UK that were supplying cars and are shifting focus to heavy vehicles. Germany is to lose a quarter of its stations by mid 2025, while Austria is losing them all. These shutdowns can only further undermine confidence, leaving prospective owners wondering if there will soon be nowhere to refuel.

Declining Support

With the possible exception of the FCEV-manufacturing Asian nations, governments around the world appear to be losing interest in hydrogen mobility.

In California, subsidies for new hydrogen filling stations were reduced in 2023, with some lawmakers pushing for their elimination entirely, arguing that uptake remains too low to justify further investment, despite nearly two decades of public funding.

The mood is the same in the US Federal government. A 2023 *National Blueprint for Transport Decarbonisation* does not envision hydrogen playing any part in light-duty vehicles by 2050. It does however have hydrogen pencilled in for extensive use in long-haul heavy-duty trucking, and in the mix along with batteries for short-haul trucks and buses.

Meanwhile in Europe, governments are being urged to take a similar path. The UK's Climate Change Committee, which provides independent advice to government, has explicitly stated that hydrogen fuel-cell cars have no significant role to play in the future of road transport and that policy should focus on battery electrification instead.

Although not an official body, Germany's Taxpayers Federation (BdSt) has publicly called on its government to halt FCEV subsidies, describing them as an "absurd use of taxpayers' money." The group estimates that at least $490 million has been spent on hydrogen vehicles and fuelling stations—around $260,000 per car sold.

Elsewhere in Europe, hydrogen strategies tend to emphasise heavy-duty transport, with little mention of personal mobility.

Taken together with collapsing sales, high running costs, and mounting infrastructure challenges, it is hard to conclude anything other than battery-electric vehicles have decisively won the contest for personal mobility. Whatever future hydrogen has in road transport will likely be limited to niche commercial applications such as long-haul heavy trucking – and only if these can outcompete rapidly improving battery alternatives on cost.

Rail

Railways have been a pillar of transportation systems since the 19th century. The advent of steam engines was revolutionary, enabling the swift movement of people and goods across vast distances. Although they now face competition from planes and automobiles, trains continue to play a pivotal role in transport, especially through electrified high speed rail. Indeed, the relative ease with which railways can be electrified positions them as an essential element of a net-zero future.

The IEA calculated that in 2022, the global average well to wheel greenhouse gas intensity for rail was just over 22 g CO_2/passenger-km. This was by far the lowest of all motorised forms of transport; half that of two-wheel vehicles and around a sixth of that from cars and planes. The IEA also estimates that if all rail journeys were replaced with road vehicles, transport emissions would increase by 12% and oil demand would go up by 16%.[2]

Having long ago set aside the steam engine, trains today are powered either by onboard diesel generators that produce electricity to drive electric motors, or by direct electrification. For climate protection, zero-emission electric trains are obviously preferred over diesel. This of course assumes the electricity is clean. If a train were to run off pure coal-fired power, the WTW emissions would be comparable to or even slightly worse than diesel.

Electricity can be continuously supplied to trains either through overhead lines, which is known as a catenary system, or via an electrified third rail that runs underneath the train. Catenary systems provide high voltage alternating current, which can be efficiently transmitted over long distances and consumed directly by electric motors. This tends to be the most efficient power system because it minimises the energy transformation steps, which is where nature's tax collector takes a cut. It also allows lighter carriages to be used, which require less energy to move.

Third rails on the other hand transmit direct current, so the trains that run on them require heavy transformers to convert the electricity back to alternating current for the motors. These extra steps reduce efficiency. The advantage of third rail is that it is more cost effective to install and does not require space above the train, which is a big advantage when tunnelling is required. They are also more robust, require less upkeep, and are less susceptible to storm damage. For these reasons they are often used in built up urban environments and underground metros.

The electrification of rail has been advancing for decades. Around three quarters of all passenger journeys are now powered by electricity, along with half of rail freight. However, global rail usage is highly concentrated in a small number of regions. Passenger rail in China, India, Japan, Russia and the European Union combines to make up around 90% of all journeys and most of these regions have electrification in excess of 80%. Other places such as Australia, Brazil, South Africa and North America have relatively low passenger volumes, but shift a lot of freight using trains that often run on diesel.

Given its energy efficiency and low carbon footprint, there is no doubt that rail powered by clean electricity will be vital to achieving a net-zero world. There is also little argument that direct electrification is the best method for powering these trains, and we should minimise the use of diesel power as much as possible.

However, there are situations where direct electrification may not be a feasible solution. The high initial costs of infrastructure such as overhead wires and substations, can be prohibitive, particularly for routes that span vast distances or pass through challenging terrain. Likewise, it can be difficult to justify the infrastructure cost for remote and rural areas with a relatively low demand. A less obvious obstacle that is encountered in some instances is ownership structure. Rail networks owned by private companies with the purpose of making money for shareholders can be reluctant to make expensive investments in electrification while the business is chugging along nicely as is. The US is a notable example, where the Association of American Railroads (AAR) whose members collectively own almost the entire network, have openly opposed direct electrification.[12]

In situations where direct electrification isn't feasible, indirect electrification is the next-best low-carbon option. As with motor vehicles, this means using onboard energy storage, either through batteries or hydrogen. The same general conclusions apply: battery-electric trains are far more energy efficient than hydrogen-powered models. Battery-electric trains typically achieve a tank-to-wheel (TTW) efficiency around 76%, while hydrogen fuel cell trains are closer to 40%.[13] Trains powered by combustion engines, whether using diesel or hydrogen, have TTW efficiencies closer to 30%. For reference, a catenary system train uses around 7% less energy per kilometre than a battery-electric train of the same weight.[14]

The experience of early adopters in Germany has been that hydrogen trains are also more expensive to run. The public transport company LNVG, which is owned by the state of Lower-Saxony began operating the world's first railway line powered exclusively by hydrogen in 2022. Fourteen hydrogen fuel-cell models were purchased to replace the diesel passenger trains running through the scenic countryside between Cuxhaven and Buxtehude. However, barely a year later, it was reported that LNVG intend to replace the remainder of their diesel fleet with 102 battery electric trains, and 27 catenary electric models because they are "cheaper to operate" than hydrogen.

The size of the cost difference can be inferred from a study by the German state of Baden-Württemberg. Assessing the decarbonisation options across 16 routes, they found that hydrogen trains would cost 70–80% more than catenary and battery trains over a 30 year period. A similar conclusion was reached in Austria. The Zillertal railway, known for offering tourist rides on an historic steam train through the picturesque 'Sound of Music' scenery of the Tyrol Valley, recently ditched plans to introduce hydrogen trains following an independent assessment by the Technical University of Vienna that recommended electrification instead. According to reporting by *Hydrogen Insight*, the chief technology officer responsible for the hydrogen proposal later lost his job after local media uncovered that he had falsely claimed to hold a doctorate for years—and that the thesis he eventually submitted was almost entirely plagiarised. A curious twist, and probably not quite the plotline Zillertal's marketing team had in mind.

In an interview with the German business magazine *Wirtschaftswoche*, Steffan Obst, head of German sales at train manufacturer Stadler—which offers both hydrogen and battery trains—said, "If the tenders are open to all technologies and the only condition is a potentially CO_2-free drive, we've found that battery almost always wins out over hydrogen."

He explains that hydrogen fuel cells require significantly more maintenance and need to be replaced every three years on average. Also, the distance between stations in central Europe is generally small, which negates the main advantage of hydrogen, which is longer range. "On most of the approximately 500 routes in Germany that are currently served with diesel, batteries are the more efficient and cheaper solution."

Once at a station, battery trains can quickly charge at high voltage, giving them enough juice to reach the next stop. However, building battery trains with a range of more than 200 km is not economical at the moment. Hydrogen trains by comparison can take on enough fuel in 15 minutes to travel 500–600 km.

So, even though batteries are preferable, there will be instances where hydrogen is selected. India plans to fully electrify its broad-gauge railways and had reached 83% at the start of 2023. However, narrow gauge rail was used in remote locations and mountainous terrain where broad gauge would have been difficult or expensive to build, and the same barriers hold for electrification. Its "Hydrogen for Heritage" programme will reportedly acquire 35 new hydrogen-engines to run along its narrow gauge 'heritage routes'. This would make Indian Railways the world's largest operator of hydrogen trains.

Similarly, the California Department of Transportation has so far ordered 10 hydrogen fuel-cell trains in an effort to decarbonise a system where rail lines are privately owned. It hopes to have them running in 2027, starting with a 160 km stretch between Merced and Sacramento. Meanwhile, the world's first conversion of a diesel train to hydrogen fuel cell was successfully completed by a Chinese state-owned manufacturer. The *Ningdong* locomotive conversion is said to be cheaper than building a fuel cell model from scratch, and can run continuously for 190 hrs. The company also says that around 90% of China's 7800 diesel locomotives are suitable for conversion, potentially making this an economically attractive decarbonisation option given China's vast distances and renewable power capacity.

Aviation

Aircraft were responsible for 633 Mt of carbon dioxide in 2019, which was 2% of global energy related emissions, as judged by the IEA. Aviation's overall contribution to global warming is thought to be even greater than this however. It has been calculated that aviation is responsible for around 4% of human-induced warming to date when emissions of other species like nitrogen oxides and water vapour are counted along with carbon dioxide.[15] Fuelled by ever-increasing demand, emissions had been growing by a little over 2% per year until a drop-off caused by the Covid 19 pandemic. We are well on our way back to business as usual however, and it's expected that a new peak will be reached in 2025.

Aircraft are of course exclusively fuelled by liquid hydrocarbons in a mixture commonly known as kerosene or jet fuel. Given that space and weight are both in limited supply on an aircraft, replacing the high energy density of jet fuel is extremely challenging. We can see just how challenging it is by comparing the densities of jet fuel with our go-to carbon-free replacements: hydrogen and batteries.

While they are already adequate for cars and emerging for trucking, a good commercially available lithium-ion battery today provides a pitifully low fraction of the energy required by a plane. A 250 Wh/kg battery holds just 2% by weight and 6% by volume of the energy contained in jet fuel. Batteries clearly have a long way to go before they are feasible for aircraft. It is thought that by 2050 we may have batteries that can hold 800 Wh/kg, which could power a 200 seat plane for at least 1000 km.[16]

Hydrogen on the other hand has almost three times more energy per kilogram than jet fuel, but as usual, is severely limited by its low volumetric energy density. For a given fuel tank, a compressed hydrogen plane would hold one seventh the energy of a conventional plane. Liquid hydrogen is of course better but still cannot manage even a quarter of the energy provided by jet fuel, and this is without accounting for the extra weight that cryogenic storage tanks require.

A comparison of jet fuel's energy density with those of hydrogen and batteries is provided in Table 7.3. We can see that regardless of the economics and technical challenges, hydrogen's physical properties mean that it cannot fully replace kerosene as aircraft fuel. However, it could be adequate for shorter flights.

This is demonstrated in a study by the International Council on Clean Transportation (ICCT), which looked at commercial flights in 2019 and analysed how many could have theoretically been fuelled by hydrogen.[17] They found that hydrogen could power a smaller regional propeller-plane with around 70 seats for 1400 km, and a 165-seat jet plane for 3400 km. This larger plane is comparable to the Boeing 737 family and the Airbus A320 family. This is the most used size of aircraft,

Table 7.3. Energy density comparison of jet fuel with hydrogen and Li-ion batteries.

	Jet Fuel (kerosene)	Compressed Hydrogen	Liquid Hydrogen	Li-Ion Batteries
Energy Density by weight (kWh/kg)	11.9	33.3 (280%)	33.3 (280%)	0.25 (2%)
Energy Density by volume (kWh/m³)	9640	1390 (14%)	2360 (24%)	625 (6%)

Table 7.4. Capacity and range comparison: hydrogen planes and commercial aircraft.

Aircraft Family	Passenger Capacity	Range (km)
ICCT Hydrogen Prop	70	1400
ICCT Hydrogen Jet	165	3400
Airbus A320/Boeing 737	160–240	6000–7000
Airbus A380/Boeing 747	Over 500	13,000–14,000

Source: ICCT (2022).[17]

with over 20 thousand currently in service around the world. These aircraft are not suitable for the longest routes, however. Long-haul flights typically use a member of the Airbus A380 family or the Boeing 747 family, for which there are a few hundred in service. The range and capacity of these craft are compared in Table 7.4 below.

It is clear that hydrogen is inadequate for the biggest planes and the longest routes. The ICCT study concluded that hydrogen could be used for between 60%–70% of flying currently serviced by the A320/737 class, however. It would also be suitable for up to 97% of flights by small regional planes. Taken all together, hydrogen could provide enough energy for around one third of all commercial flights offered in 2019. This of course means that two thirds of commercial flights are beyond its capability. Hydrogen therefore has the potential to play some role in the decarbonisation of flight, but it is far from a cure-all. Nevertheless, there are substantial developments occurring in the area of hydrogen flight.

In 2023, a German-based company named H2Fly claimed to have carried out the world's first liquid hydrogen powered piloted flight: a three hour trip in a single occupant fuel cell demonstration model. A few months later, Airbus attached a hydrogen combustion engine to a glider to complete the first flight with this form of propulsion. The purpose of the flight was to investigate contrails from hydrogen combustion however, and the glider was first towed into the air by a conventional plane.

Some companies are pushing towards near-term commercialization by developing engines that can be retrofitted into existing planes. A collaboration between budget airline EasyJet and engine maker Rolls-Royce has demonstrated that burning hydrogen can provide sufficient thrust for take-off and are continuing development with the goal of planes flying by the mid-2030s. Fellow UK-based company ZeroAvia has more short-term goals. In early 2023 they flew a 19-seat test plane with one 600 kW hydrogen engine and one conventional engine and are also developing a 2400 kW system for larger planes. They claim to have around 2000 preorders for their conversion kits and hopes to have the smaller planes flying commercially by 2025 with the larger conversions taking off a year later. However, there is scepticism that safety regulators will be ready to give them the green light this quickly.

The large manufacturers have also been involved, although interestingly, they appear to have differing opinions on the future role of hydrogen in aviation. The ZEROe program of French-based manufacturer Airbus investigated both fuel cells and hydrogen combustion engines with hopes of offering a commercial hydrogen-fuelled plane by 2035. However, in early 2025, *Reuters* reported that Airbus is suspending the programme, citing slower than expected developments in technology.

Airbus' American competitor Boeing has carried out a half dozen hydrogen projects since 2008. However, it is even more bearish than its competitor, with CEO Todd Cirton questioning the safety of using hydrogen as a fuel given that it is significantly more flammable than kerosene. A Boeing representative has told the news outlet *Hydrogen Insight* "Boeing's position is that the first, best and primary use of hydrogen in aviation should be used to develop and scale sustainable aviation fuels."

As described in Chapter 4, sustainable aviation fuels (SAFs) are hydrocarbon fuels that have been synthesised using renewable feedstocks rather than being extracted from oil and include e-kerosene and HEFA biojet fuel.

The airline industry appears to agree with Boeing on the relative contributions of hydrogen and SAF. In its net-zero roadmap, the trade association of the world's airlines (IATA), indicates that only around 4% of emission reductions by 2050 will be due to the direct use of hydrogen fuel.[18] They predict that the biggest contribution by far will come from SAF, which they expect to provide two thirds of emissions reductions.

Within the portfolio of SAFs, it is HEFA-derived biojet fuel that appears to have the biggest role. It is estimated that 90% of current SAF production is HEFA biojet, and that while other forms of SAF will increase their share in coming years, biojet fuel will still make up around 70% of SAF output in 2030.

This expectation of increased biojet production has implications for hydrogen because a good amount of it would be required. Between 20 and 45 kg of hydrogen are needed for every tonne of biojet fuel.[19] Because it's still fairly niche, biojet's total green hydrogen requirement is currently not large, although it is growing rapidly. The IATA has said that 2024 SAF production will be around 1.5 Mt. This is triple the amount produced in 2023 (480 Mt), which itself was double the 2022 figure. Assuming 90% of this is biojet fuel, we can estimate that between 30 and 70 thousand tonnes of hydrogen were needed. This is only 0.05% of global hydrogen production, however for the aviation industry to meet the target of net zero by 2050, SAF production will need to increase dramatically.

Projections of SAF annual production needed by 2050 run between 300 and 500 million tonnes, but let's use the IATA figure of 400 Mt, which is in the middle of the range.[20] This assumes that SAF will constitute two thirds of aviation fuel by this time. Supposing that 70% of SAF will be biojet fuel, this would require between 4 and 16 Mt of hydrogen – an average of 10% of current global production.

In addition to the question of green hydrogen availability, our ability to ramp up biofuel production is also limited by the availability of biological feedstock. The global production of vegetable oils – palm, soy, rapeseed, sunflower – amounts to 220 Mt per year.[20] Most of this is used for food, but even if all of it could be used for biojet fuel, it would barely cover half of the approximately 300 Mt of jet fuel used today, let alone provide for future growth.

Waste fats and cooking oils are a much more politically acceptable feedstock. Unfortunately, at 30–40 Mt per year, their volume is also a lot smaller. Assuming all waste fats and oils could be captured and processed, which is dubious, we're still only looking at 10% of current demand. Existing commercial biojet methods clearly cannot scale up sufficiently, so innovation is needed to increase our portfolio of processes, and enable the use of other feedstocks such as algae and municipal waste.

Returning to aviation's overall contribution to global warming, the latest science suggests that carbon dioxide is only responsible for about a third of its warming influence. More than 50% of the warming is now thought to be caused by condensation trails, or contrails.[21] Not to be confused with 'chemtrails', which are an invention of conspiracy-theorists, contrails are the wispy white streaks that can often be seen in a plane's wake. They are in fact ice crystals, formed when water from the jet exhaust meets cold moist air in the sky. Contrails can hang about as individual plumes for several hours or can participate in the formation of contrail cirrus clouds. These clouds of ice can then act as a blanket; trapping heat that is leaving earth on route to space.

Contrails only form under certain conditions: the air must be cold enough for ice crystals to form and humid enough for the crystals to persist. Humidity of course refers to the amount of water vapour in air. Moist air resist sublimation, which is the process where ice transforms directly into gaseous water, skipping the melting step. The sticky, uncomfortable feeling we experience in humid weather is similar – sweat evaporates less efficiently when the air has a high water content.

Ice formation is also promoted by the presence of tiny particles that give the crystals something to anchor on to. The soot that is released from jet engines fills this roll nicely.

The finding that contrails are the most significant aviation emission is notable in the context of hydrogen fuel, which produces around two and a half times more water vapour than kerosene. If hydrogen-powered aviation leads to a higher rate of contrail formation, the guilt-free flying narrative would be in danger. It is possible that the lack of soot from hydrogen engines will result in less crystal formation, but naturally occurring dust particles will still be present. Modelling suggests that hydrogen engines will produce larger crystals that could quickly drop out of the atmosphere. On the other hand, hydrogen's higher water content may mean that contrails can form under a wider range of conditions, resulting in more contrails overall.

Basically, we still don't know enough on this topic, which is why Airbus launched the project *Blue Condor*. This study involved the previously mentioned glider fitted with a hydrogen combustion engine. It also included another glider fitted with a kerosene engine. The twin gliders were set soaring under suitable conditions while a third plane loaded with analysis equipment followed with the purpose of breathing in the fumes. The goal was to measure the amounts of water, NOx, and other trace gases released, along with the size and number of ice crystals. Unfortunately, Airbus has not released the results of their study.

The outlook for decarbonizing aviation remains sobering. Liquid hydrocarbons will still be needed for long-haul flights, and the most promising low-carbon alternatives—hydrogen and SAFs—each come with serious constraints. Hydrogen may offer a solution for regional services and some medium-haul routes, but its energy density limit its broader use. SAFs, can work with existing aircraft and deliver emissions cuts over time, but their scalability is constrained by feedstock availability. Even if these fuels succeed in lowering carbon emissions, they do little to address non-CO_2 effects like contrails, which may prove even harder to avoid. As much as airlines would wish it otherwise, it seems that environmentally-conscious passengers will continue to find the prospect of flying morally problematic.

Shipping

Despite its low profile, global shipping accounts for roughly 2% of all greenhouse gas emissions. The International Maritime Organisation (IMO), which is the branch of the United Nations that oversees ships, has set goals to reduce this to near net-zero by around 2050. It also aims to see some uptake of low-carbon fuels, along with 20–30% reduction in emissions by 2030.

Both batteries and hydrogen take up too much space to be a solution for large ocean-going vessels, although they may work for smaller boats that ply rivers or coastal regions. Against this backdrop, the alternate fuels that are receiving the most attention are methanol and ammonia. A recent study by the French government-backed research institute IFP Energies Nouvelles (IFPEN) estimated that well-to-wake greenhouse gas emissions from e-methanol are 60–80% lower than those from conventional marine fuels.[22] Green ammonia offers even greater potential reductions, but with more uncertainty due to its earlier stage of development—IFPEN reports a possible range of 35–85%.

Table 7.5 compares properties of these alternatives with that of low-sulphur fuel oil: the dominant marine fuel today. As is always the case, the energy density of the alternatives is much lower than that of the fossil fuel incumbent. In order to have the same range as a ship running on fuel oil a methanol-burning ship would need a fuel tank nearly 2.5 times bigger, while the ammonia ship would need three times the storage space. This obviously means less space is available for cargo. These comparisons also assume similar engine efficiency—a generous assumption given that ammonia engines, in particular, are still in the early stages of development.

Notably, both methanol and ammonia are more than twice the price of fuel oil. It is hoped that this will drop by around 25% before 2030 due to price reductions

Table 7.5. Property comparison of fuel oil with proposed green shipping.

Property	Low-Sulphur Heavy Fuel Oil	Green Methanol	Green Ammonia
Specific Energy (kWh/kg)	11.0	5.6	5.2
Liquid Density (kg/m³)	991	798	682
Energy Density (kWh/m³)	10,900	4470	3550
Boiling Pt (°C)	> 180	65	−33.3
2024 Cost ($/MWh)*	65	170	145
2030 Cost Projection ($/MWh)*	-	125	110

Source: Lee et al. (2024).[23] Production Costs from IEA (2024).[24]

in hydrogen electrolysers and the renewable power that drives them. They will still be a lot more expensive than what shipping companies are used to paying for fuel however.

It's expected that green methanol will always be a bit more expensive than green ammonia because of the way they are made, which was described in Chapter 5. Methanol requires biogenic CO_2, which must be purchased from some other process, whereas ammonia simply requires nitrogen which can be captured from air onsite and at lower cost.

In addition to the increased fuel cost and reduction of cargo space, these alternate fuels also require vessels with more expensive engines, more safety features and higher skilled staff to operate them. The IEA estimates that once you add everything up, the total cost of ship ownership will increase by about 75%.[23] Interestingly, the cost is very much the same for methanol and ammonia. This is because ammonia is cheaper to make but more difficult and expensive to handle.

The IEA also suggests that in the case of container ships, this cost increase is not significant as long as it can be passed on to the end customers. Given the amount of cargo that these ships can carry, the increased cost would add less than one cent to the price of a smartphone or $1.50 to the price of a solar panel.[23] However, they do not mention the effect on other classes such as fuel tankers and passenger ships, whose customers may be less willing to absorb the added cost.

Nevertheless, baby steps have been taken in the low-carbon shipping transition. Methanol shipping is a little further along than ammonia. The IMO already has guidelines for the use of methanol in engines, while an equivalent for ammonia is still under development. The IMO approved initial safety guidelines in late 2024, but full regulations are still in the works, so ammonia is only being used in a few tightly controlled demonstration projects for now.

Given that large ocean-going ships take years to build, and that their work life can span decades, it is easy to see why pioneering companies that want to push on with decarbonisation are getting their orders in now – even before the engines were commercialised in some cases. *Hydrogen Insight* has reported that 166 new methanol ships were ordered in 2024, along with 27 ammonia ships. This was around a 25% increase on the previous year.

Methanol engines, and the approval to use them, were available earlier, which explains why there are a lot more methanol ships on the order books and why they are hitting the water first. Although, given the feedstock limitations of methanol and the fact that its use still emits carbon dioxide, it is expected that in the longer term, ammonia would have a bigger share of clean shipping fuel. In its net-zero scenario, the IEA assumes five times more ammonia will be used by 2030, compared to methanol. However, it remains far from certain which molecule will ultimately win out, or whether they will co-exist.

The shipping giant Maersk currently has two container ships capable of running on methanol (and conventional fuel), with another 23 reportedly on order. The smaller of the two, the Laura Maersk, took her first voyage powered by green methanol in mid 2023. The Ane Maersk became the first large ocean-going container ship to run on green methanol in early 2024.

The world's first commercially available ammonia-fuelled marine engine was launched in late 2023, and the first ocean-going vessels built to use it are expected to be completed in 2025. However, a demonstration of ammonia-fuelled shipping did take place in the interim.

In March 2024, at the Port of Singapore, the *Fortescue Green Pioneer* became the first vessel to operate on ammonia fuel. This former offshore supply vessel was converted in-house by the Australian mining-cum-green energy company Fortescue. In addition to the engineering work, the demonstration also required certification and support from the Port of Singapore to allow ammonia to be legally used as a fuel. The Green Pioneer runs on four diesel engines from the 1970s and two of them have been converted to run on ammonia with diesel acting as the pilot fuel.

In a joint press release, Fortescue and the Singapore Port Authority stated: "Post combustion NOx levels met local air quality standard, while efforts to reduce the pilot fuel for combustion ignition and N_2O emissions post combustion will continue as more ammonia-fuelled marine engines become available".

This gives the impression that an SCR system is necessary to deal with the NOx from an ammonia combustion engine, but the powerful greenhouse gas nitrous oxide, which SCR does not target, is also emitted in meaningful quantities. The Green Pioneer running on ammonia and diesel is said to produce 60% fewer emissions compared to diesel alone, and that with more modern engines this may be closer to 90%. It's worth mentioning here that the makers of the first commercially available engine – the Wärtsilä 25 – claim emissions reductions of 70–89% compared to conventional engines. So, it appears that shipping cannot be zero-emission in the same way as road vehicles, but we can still make substantial emissions reductions.

Hydrogen Insight's reporting on the Green Pioneer's conversion gives us a sense of what running an ammonia-fuelled ship entails:

The Green Pioneer stores its ammonia fuel in a deck-mounted tank said to be "literally bullet proof". Once the connections are fitted, the liquid ammonia fuelling operation is carried out remotely from the bridge to ensure no personnel are nearby in the event of an accidental release. All ammonia pipes are enclosed within a secondary containment pipe to protect against onboard leaks. The accommodation area also has refuge zones with an independent air supply so personnel have somewhere to shelter if ammonia did somehow escape. Additionally, the ship has its own Selective Catalytic Reduction (SCR) scrubbing system to treat the exhaust emissions.

These precautions give a sense of what's required to safely handle ammonia at sea—and why public acceptance could be a serious hurdle if a major accident were to occur early in the transition. While methanol is flammable and toxic at high concentrations, it dissolves readily in water and biodegrades within a week, making a spill at sea relatively benign.[25] Ammonia, on the other hand, represents a potentially more serious risk.

Because it dissolves in water, an ammonia spill would not float on the surface like oil but instead disperse throughout the water, creating a toxic environment for marine life. In contrast to a conventional shipping fuel spill, which greatly impacts marine mammals and birds but not fish, an ammonia spill could be deadly for fish with moderate impacts on birds and mammals.[26] Ammonia also promotes plant growth in water, just as it does on land. A spill would supercharge growth of algae, which leads to acidic waters that are low in oxygen. This situation is known as eutrophication and can result in aquatic dead-zones. The sea life in estuaries, mangroves and wetlands are thought to be especially vulnerable.

If ammonia were to become a widely used fuel, it would unfortunately be only a matter of time before spills occurred. In the ten years spanning 2013 to 2022, over 800 vessels were lost at sea. Nearly half of these were cargo ships, container ships, and tankers.

The risk to human life, and the negative public relations from an incident, is arguably an even greater risk. Like ammonia, methanol is toxic, albeit at higher concentrations, and also more flammable. The handling of both requires purpose designed systems and special training. At least one shipping executive has flagged the need for improved training for seafarers to avoid accidents associated with clean marine fuels. Speaking at the *World Hydrogen Summit* in 2024, Budd Darr of the Mediterranean Shipping Company likened the industry's situation to "changing the tires on the truck while it's rolling down the highway". The current training regime is not fit for purpose and transitioning to cleaner shipping fuels will require significant recruiting and upskilling of the workforce.

Along with the elevated safety risks, is a more fundamental challenge for these would-be fuels – having enough of them. In 2022, the international shipping industry used about 2,600 TWh of fuel, almost all of it fossil-based. Replacing this with alternative fuels would require around 500 million tonnes of ammonia or 460 million tonnes of methanol per year.

Current global production falls far short of that. Ammonia is currently made at a rate of 180 million tonnes per year, while methanol is closer to 90 million tonnes.

What's more, nearly all of this is made from fossil gas so using it as fuel would potentially be worse for the climate than heavy fuel-oil. Ramping up production of low-carbon ammonia and methanol will be a mammoth task.

Producing 500 million tonnes of green ammonia would require about 88 Mt of green hydrogen, while 460 million tonnes of green methanol needs about 92 Mt. In either case, 4500–5000 TWh of renewable electricity would be required to make the green hydrogen. For context, total global renewable electricity generation in 2022 was a little over 8,500 TWh—and wind and solar contributed just 3,400 TWh of that. Clearly, we would need to build a great deal more clean power.

However, this could be done with time. In 2023, a record 500 GW of renewable power generation capacity was added globally, mostly as solar and wind in China.[27] If we assume that the sun shines and the wind blows somewhere between 20 and 25% of the time, then this new capacity provides 880 to 1100 TWh of power per year. That's 10–12% more renewable electricity from just one year of building.

If new renewable generation continues to build at this rate, it would take between 5 and 6 years to have enough new capacity to fuel all international shipping at 2023 levels. Of course, shipping is expected to grow and its energy requirement along with it. Also, shipping doesn't get to hog all the new energy. There are many other sectors that will be lining up for their share as well. However, the numbers do suggest that the scale of the challenge is not out of the question over a timespan of decades.

In addition to renewable generation, we will also need to build the electrolysers to convert that electricity into the hydrogen needed to make the fuels. In the case of ammonia, there is also much storage and handling infrastructure required, not to mention building of the ships themselves.

A research group from the University of Oxford has crunched numbers to estimate how much all this building might cost.[28] They found that production and distribution of green ammonia in quantities 3 to 4 times greater than the fossil ammonia that we use today would cost around $2 trillion by 2050; $80 billion per year over 25 years. This figure includes building out the renewable electricity generation, hydrogen and ammonia production facilities, pipelines and port storage facilities. It does not include ships.

These are no doubt big numbers, but for context, the cost of the 230 million tonnes of fuel that the shipping industry used in 2023 at an average price around $700/t would have totalled $160 billion. These calculations suggest that while it won't be easy, we do have reasons to be optimistic about the possibility of fuelling our ships with liquids produced from renewable electricity in the coming decades.

Big Numbers: Chapter 7

- **14%** - Share of global greenhouse emissions from transport. Around 10% comes from the road vehicles, with shipping and aviation contributing 2% each.

- **2.5x** - Amount of renewable electricity required by hydrogen fuel cell vehicles to deliver the same service as battery electric vehicles—with running costs increased by a similar factor.

- **320:1** - Ratio of battery electric vehicles to hydrogen fuel cell vehicles on the world's roads in 2023. The gap continues to widen.

- **200 km** - Economic range limit for battery electric trains. For longer distances where overhead electrification isn't viable, hydrogen trains may be considered.

- **66%** - Share of commercial flights in 2019 that hydrogen could not have powered, due to its low volumetric energy density.

- **10 Mt** - Hydrogen required if all aviation in 2023 had been fuelled with sustainable aviation fuel. This would have consumed 7% of global renewable electricity generation at that time.

- **80%** - Potential reduction in well-to-wake emissions from shipping when fossil fuels are replaced with green methanol or green ammonia.

- **90 Mt** - Hydrogen required if all shipping in 2023 had used green methanol or ammonia. This would have consumed 60% of global renewable electricity generation at that time.

CHAPTER 8
The Scale Challenge

◇◇◇

Triage

Many uses for hydrogen have been proposed. Some offer deep emissions cuts. Others are a wolf in sheep's clothing – more polluting than what they would replace. All are expensive, because hydrogen itself is expensive.

In every case, technology must improve, and the cost of production must come down if they are to move from technically feasible to practical. If we choose optimism and assume this will be so, what then is hydrogen's role?

We can sort the possibilities into three broad buckets: *worth doing, worth watching, and walk away.*

Worth Doing

These are the applications where no direct electric options exist, or where hydrogen is essential. Note that they all use hydrogen for its molecular properties, rather than as a fuel directly.

- *Refining*: excluding the two thirds of consumption that is byproduct, replacing the merchant grey hydrogen used in oil refining can cut emissions without system redesign.
- *Ammonia*: there is limited opportunity within existing plant, but new plant designed to integrate green hydrogen offer true decarbonization of this important chemical.
- *Methanol*: the argument for ammonia applies here too. Bio-methanol may be cheaper in some contexts, but green hydrogen-derived methanol can still play a role.
- *Steel*: hydrogen-based DRI can cut emissions by 90%+ and is one of the most promising new uses.
- *Shipping fuel*: hydrogen-derived ammonia and methanol may be the only scalable zero-carbon options for long-haul maritime.
- *Jet fuel (SAF)*: synthetic hydrocarbon fuels made using green hydrogen are among the few long-range options available.

DOI: 10.1201/9781003361428-10

Worth Watching

Marginal applications that use hydrogen as a fuel. Their value is situation or location dependent.

- *Long-duration energy storage*: round-trip losses are high but may complement renewable generation in locations with access to suitable underground geology.
- *High-temperature industrial heat*: highly process dependent and reserved only for those that require very high temperatures and can deal with nitrogen compounds.
- *Heavy-duty, long-distance trucking*: if they can out-compete electric models, which is still uncertain.
- *Hydrogen trains*: reserved for routes that are too long or too difficult to electrify.

Walk Away

These uses fail on efficiency, cost, and basic logic—they're distractions from better solutions.

- *Power generation*: except for seasonal storage, electricity to hydrogen to electricity is a wildly inefficient use of clean energy.
- *Passenger cars*: Use 2.5–3x the primary energy of BEVs, which are a joy to own.
- *Building heat*: heat pumps are 4–6x more efficient, safer, and already established.

Our examination of the various uses that have been proposed shows that hydrogen is not a universal solution. Rather, it is an enabler for specific decarbonization challenges. Each of these applications require enormous quantities of energy and will require an unprecedented buildout of infrastructure. To get a feel for the scale of this challenge, let's do some back of the envelope calculations.

The Big Buildout

Let us assume that electrolyzers operate with an efficiency of 50 TWh/t H_2. This is a nice round number, and it sits in the middle of the range of what's currently achievable. As we gain experience with this technology, efficiencies will probably improve, but not enough to change the magnitude of our calculations.

Let us also assume that, for each application, green hydrogen is used exclusively to meet the full demand – so all steel is made using DRI, all ships are fueled by ammonia and so on. To be clear, nobody is suggesting that this will be the case anytime soon, but it provides us with an upper bound and a sense of how much energy each application draws.

Using figures from 2023, Table 8.1 lists the electricity consumption of major economies alongside approximations of the electricity that would be needed to supply the "Worth Doing" applications with green hydrogen. The 'Hydrogen Requirement' ratio for each application is either the theoretical value from chemistry, or a representative number from literature.

Table 8.1. Electricity demand of hypothetical hydrogen-route decarbonization by sector.

Use Case (Country)	2023 Global Consumption	Hydrogen Requirement	Approximate Hydrogen Quantity (Mt)	Electricity Required (TWh)
World				*29,700*
2050 H$_2$ Estimate			400	20,000
China				*9440*
Steel	1,900 Mt	60 kg/t steel	110	5500
Shipping Fuel (ammonia)	500 Mt (eq 230 Mt VLSFO)	16.2 kg/t NH$_3$	90	4500
USA				*4270*
India				*1960*
Ammonia	185 Mt	16.2 kg/t NH$_3$	30	1500
Russia				*1160*
Japan				*1013*
Methanol	90 Mt	189 kg/t CH$_3$OH	20	1000
Germany				*506*
Aviation Fuel (HEFA biojet)	300 Mt	33 kg/t biojet	10	500

Source: National electricity consumption data from Ember (2023).

The result is eye-opening. Producing jet fuel with green hydrogen would require as much electricity as Germany consumes in a year. Methanol would match Japan. Ammonia and steel would exceed every nation but China. And the 400 Mt of hydrogen projected under net-zero scenarios (as discussed in Chapter 1) would demand electricity equivalent to two-thirds of global consumption in 2023.

These figures speak for themselves — it's a staggering amount of electricity. To get a feel for what this means in terms of infrastructure, let's do a little more arithmetic.

Electrolyzer Buildout

The cost of green hydrogen depends on several factors, but the most important are the electricity price and the amount of time that electrolyzer sit idle. IRENA has reported that electrolyzers running for 3000–4000 hours per year can make economically viable hydrogen, but hybrid plants at the best locations in the world incorporating both wind and solar, could achieve capacity factors above 5000 hours.[1] So, let's assume that our electrolyzers are in a great location and can run for 5000 hours a year.

Using this assumption, we can estimate the electrolyzer capacity required to produce 400 Mt of hydrogen in 2050 with the following calculation:

$$400 \text{ Mt H}_2 \times 50 \text{ TWh/Mt H}_2 \div 5000 \text{ h} = 4 \text{ TW}$$

Electrolyzer stacks are typically available in the 1–5 MW size range, so 4 TW would require 800,000 of the largest. According to the IEA, the global installed electrolyzer capacity at the end of 2023 was 1.4 GW – a tiny fraction of the 2050 requirement.

To reach 4 TW by 2050, we would need to increase installed capacity by over 2800 times, which is around a 38% increase every year – an average of 148 GW/yr. This is a highly challenging prospect, but not outside the realm of possibility. According to the International Solar Alliance, solar capacity has expanded at an annual rate of 37% since the year 2000, offering an optimistic point of comparison.[2]

Achieving this buildout will require more electrolyzer factories. The IEA has reported that as of 2023, we can build about 25 GW of new electrolyzers in a year. Planned expansions could increase this to 165 GW by 2030, which would see us more or less on track – if it happens. However, expansion is something of a tightrope walk for manufacturers. Expand too quickly and they risk being left with insufficient buyers and idle equipment; wait too long and they risk missing out when demand surges. Getting this wrong is hazardous to business health.

Renewable Power and Land Requirements

Aside from money, electrolyzer expansion is only constrained by our ability to build and install them. The space they take up is insignificant. In contrast, the renewables that power them require vast tracts of land, in addition to generation equipment. To estimate how much land this might require, we'll take a slightly different approach and use the spatial energy density metric, which expresses how much electricity a technology can provide per unit of land it takes up.

While hybrid wind–solar farms are still rare, a global study suggests they generate about 50% more electricity than solar alone. Utility-scale solar typically delivers around 0.065 TWh/km², so a hybrid setup can be expected to reach roughly 0.1 TWh/km².[3]

The NEOM green hydrogen facility being built in Saudi Arabia includes approximately 4 GW of wind and solar generation and covers an area of 300 km². Assuming that most of this site is taken up with windmills and solar panels, and that they average 36% of maximum output over time,[4] the energy density of this project is roughly 0.04 TWh/km². As the first large scale green hydrogen facility of its kind, this is a useful benchmark for economic viability.

Including both figures provides a range for the land that will be required. Table 8.1 shows that our 400 Mt of hydrogen requires around 20,000 TWh so we can calculate the space it requires as follows:

$$20,000 \text{ TWh} \div 0.04 \text{ TWh/km}^2 = 500,000 \text{ km}^2$$

and

$$20,000 \text{ TWh} \div 0.1 \text{ TWh/km}^2 = 200,000 \text{ km}^2$$

This is a large area. Approximately 140 of the world's nations are less than 200,000 km² in size. However, using a lot of space is a defining feature of humanity.

According to *Our World in Data*, around 50 million km², or half of the world's habitable land is taken up with agriculture. In fact, 5 million km² is used just for things we don't eat, like cotton and biofuel. In this context, 500,000 km² for green hydrogen is quite modest. It becomes even less demanding when considering that renewables can be sited on land with limited agricultural value, can co-exist with farming operations, and can be deployed offshore.

Moreover, it would represent only a fraction of our total renewable generation capacity, assuming electrification progresses as required by low-emissions development pathways.

Using the IEA's *Net Zero Scenario* as one example, the annual renewable electricity supply would increase from 12,000 TWh in 2023 to 86,000 TWh in 2050.[5] Our extreme case for green hydrogen's electricity requirement is less than a quarter of this.

In the *Net Zero Scenario*, about two thirds of the total electricity comes from solar and wind. This equates to 58,000 TWh, which is up from 4000 TWh in 2023. By applying the same spatial energy density figures as before (0.04–0.1 TWh/km²) we can estimate that the total additional solar and wind will take up between 0.5 and 1.35 million square kilometers.

This is still a lot less than the 5 million km² currently used for non-food crops. Based on this estimation, space should not be a limitation in absolute terms. However, the best locations for making green hydrogen derivative are often not the same as the locations that want to use them.

A 2023 study published in Nature Communications found that only about 50% of global hydrogen demand could be met by local production in regions without water or land constraints.[6] The countries with no significant land or water constraints include Australia, Argentina, Canada, Mongolia, and Namibia. Perhaps surprisingly, some very large countries face greater constraints due to much of their land being forested or having existing agricultural commitments. These include Brazil, China, Russia, and the United States. Unsurprisingly, nations that face both land and water scarcity include those of Western Europe, Japan, and South Korea – the principal hydrogen buyers.

Project Costs

The cost of building large process plants is influenced by several factors. Location plays a major role, as the rich nations of Europe, North America and Australasia tend to pay workers more and have strict environmental and safety regulations. While these measures are important for minimizing ecological impact and ensuring worker protection, they can increase the cost of industrial projects. The other side of the coin is that developments in rich countries generally pay lower interest rates on the money they borrow to build. This is because lenders view rich countries as lower in risk. The opposite is true in many nations across Africa, Asia, and Latin America, where internal costs are lower, but interest rates are significantly higher. Other factors that affect the build price include the size of the project, its location in relation to existing infrastructure, and the availability of government support.

Table 8.2. Reported production cost estimates for green hydrogen projects.

Project Name	Location	Hydrogen Production (tonnes per year)	Reported Cost ($)	Unit Cost ($/Mtpa)
NEOM	Saudi Arabia	219,000	8.4 billion	38 billion
Australian Renewable Energy Hub	Australia	1,800,000	40 billion	22 billion
Aman	Mauritania	1,700,000	40 billion	24 billion
Hyrasia One	Kazakhstan	2,000,000	40–50 billion	23 billion

Source: Hydrogen Insight

Given the above, we can only make a very rough estimate of the amount of money that will be required to build out enough production capacity to provide 400 Mtpa. As of 2025, there are no completed GW-scale projects to use as a benchmark, but we can use announced projects for which cost estimates have been reported. From Table 8.2 we can see that the projected cost for a plant with electrolyzer capacity in the neighborhood of 2 GW, along with the renewables to power it, is around $40 billion. The 2 GW mark is typical of large-scale project proposals. A few much larger projects have been announced, but these should be viewed skeptically.

It is notable that the cost per unit of hydrogen production capacity at the NEOM plant is significantly higher than the estimates for other project proposals. However, when it was originally announced in 2020, the expected cost of NEOM was only $5 billion. This translates to a unit cost of $23 billion, putting it on par with the others. It would not be surprising if the cost of other projects, should they actually get built, also escalates significantly. In any case, based on this information it is reasonable to use a range of $20–40 billion per million tonnes of capacity as a ballpark estimate.

Using these figures, a global production capacity of 400 Mtpa would cost somewhere between $8 and $16 trillion dollars.

Those are undeniably large numbers, although they would be distributed globally and spread over decades. For perspective, Japan's GDP in 2023 was $7.5 trillion, while the U.S. federal budget stood at approximately $6.3 trillion.

A more direct comparison might be the oil and gas industry's capital expenditures, which averaged $410 billion per year from 2015 to 2022. Of course, oil and gas currently supply nearly 60% of the world's energy, whereas hydrogen is projected to contribute only 10–15% by 2050—so the motivation to invest at that scale may not be comparable. Still, at that level of spending, building out the necessary green hydrogen infrastructure would take between 19 and 39 years. This suggests that achieving the 2050 target is plausible—if costs remain at the lower end of projections and companies find sufficient economic justification to proceed.

The buildout isn't simply a matter of throwing money at it, however. We also need time. Large clean hydrogen facilities have long development timelines. Typically, they require around six years for planning, permitting, design, and securing finance. This is followed by four years of construction, making it unlikely that any project not already well into development today will be operational by 2030, which is the target year for many national goals.

The Hydrogen Council reports that 48 Mtpa of hydrogen production has been announced for completion by 2030. However, rising costs and delays in implementing the US and European incentives have led to stalled projects and cancellations, reducing the expected global output to between 12 and 18 Mtpa.[7] Given that the US and EU each have clean hydrogen production targets of 10 Mtpa by 2030, it is clear that many countries goals will not be met.

In Part 3, we'll look more closely at this gap between ambition and action, and how the nations of the world are attempting to bridge it.

Big Numbers: Chapter 8

- **20,000 terawatt hours** - Estimated electricity required to produce 400 Mt of green hydrogen - two thirds of global generation in 2023.

- **800,000** - Estimated number of electrolyzers needed to supply 400 Mt of green hydrogen per year, assuming they are 5 MW each.

- **500,000 km²** - Amount of space that may be required to generate enough renewable electricity to make 400 Mt of green hydrogen per year. This is roughly the size of Spain.

PART 3

Direction: Ambitions and Actions from Around the World

CHAPTER 9
Making the Rules

◇◇◇

Japan became the first nation to publish a formal hydrogen strategy in 2017. For a time, it stood alone – an island in both geography and policy – waiting for other countries to follow. As hydrogen hype grew, so too did government interest. This is illustrated by Figure 9.1, which shows how the number of national hydrogen strategies grew over the first half of the 2020s. By the beginning of 2025, 60 countries had published a strategy or road map. According to the IEA, these nations account for 84% of global energy related carbon dioxide emissions.[1] This growing momentum was also reflected in the 2023 UN climate conference in Dubai (COP28), where, for the first time, a global climate agreement explicitly called for accelerating the use of clean hydrogen. Nowadays, when lawmakers think about the energy transition, hydrogen is clearly in the mix.

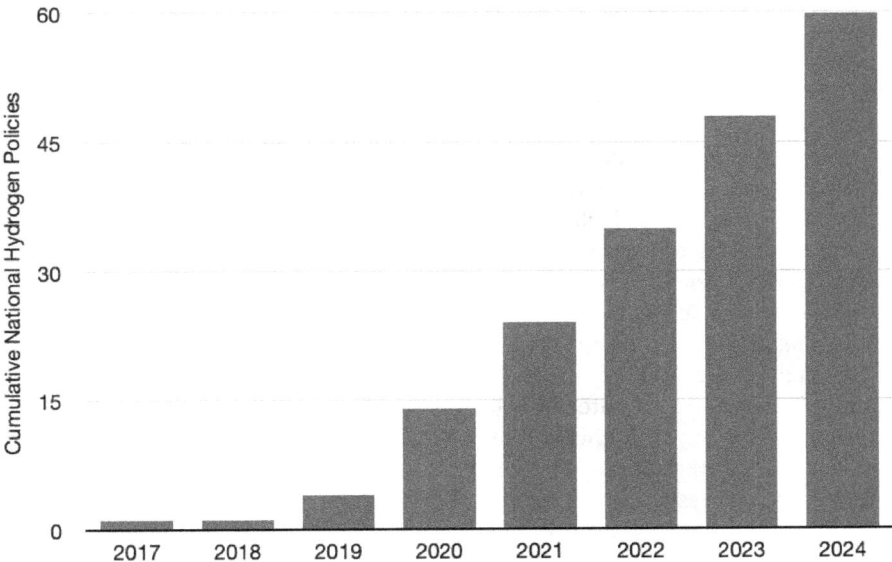

Figure 9.1. Cumulative National Hydrogen Policies by Year. Source: UNIDO, IRENA, and IDOS (2023).[2]

DOI: 10.1201/9781003361428-12

The scope and ambition inherent in hydrogen strategies does vary, however. For some governments, adopting a hydrogen roadmap appears to have been driven by a desire to avoid being seen as a laggard. Others approach it more earnestly, backing their plans with substantial funding and detail. There are a variety of motivations propelling governments to pursue hydrogen. In certain cases, it was seen as helpful in accelerating economic recovery following the Covid-19 pandemic, but longer-term motives also play a role. These include enhancing energy security, securing energy independence, reducing greenhouse gas emissions, fostering new industries and exports, and protecting existing ones. While all these objectives have universal appeal, the emphasis placed on them does vary between nations.

Advanced economies with limited access to cheap renewable energy tend to focus on maintaining and decarbonizing their industries. Their relative lack of resources has also prompted efforts to develop clean hydrogen production overseas and promote international hydrogen trade. We have labelled this group the *buyers*. It includes Japan, South Korea, and the European Union, with Germany playing a particularly prominent role. These countries are arguably the driving force behind the hydrogen economy, creating the demand that is propelling the movement forward.

On the other side of the coin are nations with abundant renewable energy resources that appear to be primarily motivated by export dollars. This group includes high income countries such as Australia, Canada, and nations of the Middle East – seeking to diversify and future-proof their energy exports. It also includes developing economies that may or may not have a history of energy exports, but see an opportunity to improve their economic prospects through renewables. This second cohort, which is largely located in Africa and Latin America, is also part of the group that we have termed the *sellers*.

Finally, there is a group of nations taking a more balanced approach to clean hydrogen – prioritizing domestic production and consumption to support decarbonization and diversification of their energy mix. We refer to these as the *DIY* (Do It Yourself) group. These nations include economic powerhouses like the United States and China, which have vast industrial bases and the energy resources to sustain them with domestic hydrogen production. The UK, though smaller in scale, also appears well-positioned to meet most of its own needs without significant reliance on imports. Also included is India: a slumbering giant with the potential to become a major hydrogen producer, driven by its expanding industrial sector and strong renewable energy potential. These countries prioritize hydrogen as a tool for domestic aims while staying open to the potential for export opportunities should a market materialise.

Broadly, the economic problem that everyone is trying to address is the difference in price between clean hydrogen, which is very high, and fossil fuels, which are very cheap. Few buyers are willing to pay more for energy, so without government intervention, a market for clean hydrogen will simply not develop. There are a range of policy approaches that counties may select from to help bridge this gap.

The first is to reduce the price that buyers pay for hydrogen by using taxpayer money to cover some of the cost of production. This may be done by direct grants for the building of hydrogen plants, which reduces the developer's building cost and consequently lowers the sales price needed to recoup the investment. Alternatively, governments can offer a subsidy payment for every kilogram of hydrogen. In theory this payment could be offered to the buyer or the producer, but most territories tend to focus on the later.

A second approach is to make fossil fuels more expensive by placing a tax on their use. This strategy is not specific to hydrogen promotion, but can be used to incentivise decarbonization more broadly. However, it is politically tricky to implement because our economies have been built on fossil fuels, and (with one notable exception) governments concerned with maintaining power are reluctant to introduce laws that increase their constituents' cost of living.

Thirdly, governments can apply mandates that require businesses to use a certain proportion of hydrogen in their energy mix, and levy fines on those that don't. In practice, some mix of all of these financial measures will likely work best.

A fourth strategy, used by many countries, is the promotion of 'hydrogen hubs'. These are regional networks where hydrogen producers and consumers are located in close geographical proximity. This helps reduce the cost of hydrogen supply by minimising transportation distances, making pipelines more feasible and less costly.

All these interventions are made with the assumption financial support will not be required forever, just long enough for the industry to find its feet – which may take decades. Many believe that the cost of clean hydrogen production will come down over time according to Wright's Law, which says that the cost of a technology reduces as its cumulative production increases. This is mostly due to people learning how to do things better and on a larger scale over time. By way of example, the cost of onshore wind power has fallen by 70% over the last decade, while the cost of PV solar and batteries have dropped by 90% or more.[3] Assuming this principle applies to green hydrogen as well, the argument goes that in 20 to 30 years' time, subsidies will no longer be required. A counterpoint might be that subsidies can be politically difficult to remove once established. After all, the world still spends hundreds of billions of dollars every year on subsidising fossil fuels.

Tracking by the energy consultancy Bloomberg NEF (Table 9.1) shows that announced public spending for the development of clean hydrogen industries was in excess of $360 billion by the middle of 2024. It's a substantial sum—though worth keeping in perspective. Much of it is spread over a decade or more, meaning annual hydrogen subsidies would amount to tens of billions rather than hundreds. For comparison, governments spent around $70 billion in 2023 on direct support for consumers to adopt clean energy technologies such as EVs, heat pumps, and insulation, while fossil fuels received $620 billion in subsidies.[4]

The second common challenge that governments face is ensuring that in addition to making economic sense, hydrogen policies have climate integrity. The rules should result in the lowering of total emissions, which after all is the main reason we are looking at hydrogen-based solutions in the first place. This begs the question – what makes hydrogen *clean*?

The standard approach taken by policymakers is to define clean hydrogen based on the greenhouse gas emissions form production measured on a well-to-gate basis. These thresholds vary by country, as shown in Table 9.2 below. Note that

Table 9.1. Summary of hydrogen funding by region as of May 2024.

	The Americas	Asia Pacific	Europe, Middle East, and North Africa
Government Funding	$188 billion -mostly from the US	$32.5 billion -mostly from Japan	$140 billion at least -mostly from Europe
Demand Quotas	None	South Korea has quotas of questionable utility	EU has the strongest quotas
Carbon Pricing	Absent or insufficient	Absent or insufficient	EU and UK have price mechanisms and plans to strengthen them

Source: BNEF (2024).[5]

Table 9.2. Clean hydrogen definitions of selected territories.

	Terms	Thresholds (kg CO_{2-eq}/kg H_2)
Brazil	Low-carbon hydrogen	≤ 7.0
Canada	Clean hydrogen	≤ 0.75 (highest tier) < 4.0 (lowest tier)
European Union	Low-carbon hydrogen Renewable hydrogen	≤ 2.8 all energy sources permitted only renewable energy sources permitted
India	Green hydrogen	≤ 2.0
Japan	Clean hydrogen	≤ 3.4 (hydrogen) ≤ 0.84 kg CO_2/kg NH_3 (ammonia)
Korea	Clean hydrogen	≤ 0.1 (highest tier) 2.0–4.0 (lowest tier)
United Kingdom	Low-carbon hydrogen	≤ 2.4
United States	Clean hydrogen	≤ 0.45 (highest tier) 2.5–4.0 (lowest tier)

there is a lack of alignment between definitions. This poses a potential challenge for international trade: what qualifies as "clean" in one jurisdiction may not make the cut in another. Some countries have also set a range of emissions limits that qualify for different levels of subsidies. The cleaner the hydrogen, the larger the financial support.

Governments may also place restrictions around allowable energy sources. For example, the EU definition of 'renewable' hydrogen excludes the blue and pink variants, whereas 'low-carbon' hydrogen allows all energy sources.

Along with emission limits, policies aimed at incentivising green hydrogen may also include a set of criteria that act as guardrails against unintentional emissions. Pioneered by the EU and subsequently adopted by the US, these guardrails are widely known as the 'three pillars' and help to ensure the climate integrity of hydrogen projects.

Pillar 1: Additionality

Renewable electricity used for hydrogen production must come from newly built generation, not existing. The concern behind this requirement is that if electricity-hungry hydrogen production were allowed to gobble up established renewables, then this electricity would need to be replaced by other generation sources to meet existing needs. These additional sources would most likely be fossil fuels, and this would drive up total emissions from the electricity sector. Additionality ensures that this does not happen by requiring the construction of *additional* renewable energy infrastructure to covers the demands of hydrogen electrolysis.

Pillar 2: Geographical Matching

The additional renewable electricity must be built in the same region as the hydrogen facility. This helps maintain grid stability and ensures that the renewable energy is available to the hydrogen production site. Without this requirement, a developer could theoretically build renewable generation in a faraway location that is not physically connected to the hydrogen plant, which would make a mockery of the additionality rule. Geographical matching ensures its integrity.

Pillar 3: Temporal Matching

Electrolysers should operate only when renewable power is available. If they run when renewables aren't generating, the electricity is likely to come from fossil fuels, making the hydrogen far from green. Temporal matching prevents this.

However, it is technically challenging and expensive to track renewable generation and match its consumption in real time. In fact, it is currently impractical, so lawmakers must select a longer time scale in which temporal matching is to occur. A strict requirement would see plants match hydrogen production with renewable power generation on an hourly basis. This method is considered to have the highest climate integrity. A more flexible regime, albeit with questionable climate credentials, would allow matching on an annual basis. In this case, hydrogen could still be produced using grid electricity when renewable generation is low, as long as the company producing the hydrogen was able to send the same amount of renewable power back to the grid at a later time.

Many hydrogen developers and industry groups prefer weak guardrails and in particular, a longer time-matching window. They argue that stricter requirements raise costs, reducing the likelihood that green hydrogen will take off, and slowing the energy transition. This is because electrolysers are most efficient and long-lived when they can operate steadily and continuously. In addition, the cost of electrolytic hydrogen is directly related to the amount of time electrolysers are running, which is known as the 'capacity factor'. The cost to build the equipment is the same whether it is turned on or not, and this cost must be recouped by selling the hydrogen it produces. The smaller the amount produced, the greater the price that must be charged to recoup costs.

On the other hand, environmental groups and some green hydrogen developers argue that strong guardrails are required to avoid scenarios in which the taxpayer is subsidising greenwashed hydrogen projects that do not provide climate benefits. They also argue that it is possible to produce hydrogen profitably under a strict three pillars regime, pointing to the low cost of electricity at times of high renewable generation. Along with the capacity factor, electricity is a major contributor to the cost of green hydrogen, and it is argued that bargain prices during times of excess generation will make up for lower capacity factors.

The tension between the pro-development and the pro-environment camps is basically a disagreement about whether we should prioritise large amounts of hydrogen with questionable cleanliness, or lesser amounts of the cleanest hydrogen. Quality versus quantity. Both sides are able to present energy modelling studies that back up their view, as the results of these studies are very much dependent on the assumptions made. However, it is perhaps telling that those promoting weaker guardrails are mostly those with a financial interest in hydrogen production, while those demanding strong guardrails are primarily motivated by environmental protection.

These competing pressures have made the three-pillar framework a point of political friction in both Europe and the United States, where governments have struggled to reconcile climate ambitions with market realities. As a result, the rules around additionality and time matching have become heavily contested, with calls for delays or revisions. More on this later.

It's worth noting that a broad and somewhat obvious conclusion of the energy modelling is that the cleaner the electricity grid, the less important the three pillars are. As the proportion of renewables in a nation's electricity generation mix increases, the difference between hourly and annual matching diminishes. The balance between cost and emissions largely depends on how much renewable energy is available in a given location and how quickly that supply can grow. Policy often reflects the expectation that equipment costs will come down in the future and that the proportion of renewable power in the grid will increase. Consequently, laws tend to start off relatively lenient and become increasingly strict over time, hopefully striking a balance between economy and environment.

Of course, laws hinge on the outlook of the government of the day. The third major challenge for hydrogen policymakers, along with lowering cost and maintaining environmental integrity, is the transience of government itself.

At its heart, hydrogen policy is climate policy, and not all corners of the political spectrum are sympathetic to climate concerns. In fact, some are openly hostile. Research has shown that the presence of right-wing populists in government is associated with a significant weakening of climate policy, especially in countries where the winning party governs alone. These parties, and many of their supporters, often view climate change with suspicion and see climate action as something pushed by out-of-touch elites, driven by international bodies rather than national governments, and burdening ordinary people with the costs. This kind of politics has been gaining traction around the world, making it increasingly likely to shape governments. The prospects of hydrogen can therefore rise and fall with a nation's broader commitment to climate.

This creates a challenge for developers. Planning, designing, and constructing a large-scale hydrogen facility typically takes close to a decade – far longer than the three or four-year terms of liberal democracies. This makes hydrogen projects vulnerable to political turnover. By the time developers are ready to break out the shovels, the rules may have changed.

The following outline of hydrogen policies around the world is therefore offered as a snapshot of where things stood in the middle of the 2020s. It is not exhaustive, and will no doubt change as world events, technologies, and politics themselves continue to evolve.

Big Numbers: Chapter 9

- **60** - Countries with a national hydrogen strategy by 2025, covering 84% of global energy-related CO_2 emissions. Hydrogen is now embedded in global energy and climate policy.
- **$360 billion** - Announced public funding for clean hydrogen, as of mid 2024.
- **3** - Guardrails to ensure climate integrity for green hydrogen production: additionality, temporal matching, and geographic matching.

Chapter 10

The Buyers

◇◇◇

When it comes to championing hydrogen, the European Union, Japan, and the Republic of Korea are out in front – waving the pom-poms for the hydrogen economy. These highly developed, energy-hungry players have all reached a similar conclusion: they cannot meet their clean energy needs with domestic renewables alone. Imports of clean hydrogen, they believe, must fill the gap.

Yet while their motivations overlap, their strategies differ in important ways. Table 10.1 below summarises the key elements of their approaches, which we'll explore in more detail in the sections that follow.

Table 10.1. Summary of hydrogen policies for the buyer nations.

	Strategy Document	Key Policy Instruments	Notable Targets
Japan	Basic Hydrogen Strategy (2017), updated 2023	- Production Subsidies - Direct Investment - Global Outreach	By 2030: - 3 Mtpa consumption By 2040: - 12 Mtpa consumption By 2050: - 20 Mtpa consumption - 10% electricity from NH_3/H_2
South Korea	Hydrogen Economy Roadmap (2019), updated 2022	- FCEV Subsidies - R&D Tax Credits - Global Outreach - Consumption Mandates (power generation)	By 2030 - 300,000 FCEVs - 2.4% electricity from NH_3/H_2 (5.5% by 2038)
European Union	EU Hydrogen Strategy (2020)	- Production Subsidies - Direct Investment - Global Outreach - Consumption Mandates (industry, transport)	By 2030: - 10 Mtpa production - 10 Mtpa imported

DOI: 10.1201/9781003361428-13

Team Europe

'Team Europe' is a term that was coined by the European Union (EU) during the Covid19 pandemic and loosely describes a state of organisation that involves individual member countries working in a coordinated manner alongside the EU government. The EU itself is a political and economic union of 27 European countries that collaborate on issues such as trade, security, and environmental policy, with the aim of promoting economic integration, peace, and stability across the continent.

Sadly, the aspirations for peace and stability were issued a significant challenge when Russia invaded Ukraine in February of 2022. Although Ukraine was not a member of the EU, the prospect of having borders within the continent redrawn by force is seen as intolerable. The EU found itself in a predicament. On one hand, it was obliged to impose sanctions on Russia in response to this aggression. On the other, it was reliant on Russian exports for almost a quarter of its energy needs. In 2020, Russia provided Europe with 19% of its coal, 26% of its oil and over 40% of its natural gas.[1]

Part of Europe's response to this situation was the RePowerEU plan, which was launched a month after the invasion, to accelerate the transition away from Russian fossil fuel and towards clean energy sources. RePowerEU is a non-binding, three-pronged strategy of energy efficiency, diversification of energy supplies, and clean energy production. Hydrogen is seen as a key component. RePowerEU ushered in the EU's current goal of producing 10 Mtpa and importing a further 10 Mtpa of renewable hydrogen by 2030.

However, renewable hydrogen was on the menu before Putin sent the tanks rolling in. The EU was an early adopter of hydrogen policy, releasing its first strategy in 2020 alongside the European Green Deal – its overarching roadmap for climate action. Framing itself as a global leader on environmental issues, the EU has set the ambitious goal of becoming the world's first carbon-neutral continent. Together, these initiatives command €1.8 trillion ($1.95 trillion): a third of the EU's 2021–2027 budget. Part of this sweeping agenda, the EU's hydrogen portfolio stands out as arguably the most ambitious and comprehensive in the world. Let's see what it involves.

Mandates

A key component of the Green Deal is the so-called 'Fit for 55' package. It incorporates 14 different policy initiatives that target the EU's greenhouse gas emissions, with the goal of seeing 2050's emissions 55% lower than they were in 1990. These include legally binding mandates for the use of hydrogen and its derivatives across multiple sectors—an approach that, so far, remains unique to the EU. A summary of these mandates is provided in Table 10.2 below.

Another important hydrogen-relevant element of Fit for 55 is the overhaul of the EU's Emissions Trading System (ETS). Established in 2005, the ETS was the world's first international carbon trading scheme. It operates as a 'cap and trade' system, whereby a limit is placed on the total greenhouse gas emissions allowed for targeted sectors, and companies within those sectors must acquire the right to

Table 10.2. Summary of European Union hydrogen mandates.

Policy Name	Policy Focus	Hydrogen Provisions
Renewable Energy Directive (RED3)	Legally binding renewable energy consumption targets	- 42% of hydrogen used in industry must be renewable by 2030 (60% by 2035) - 1% of transport fuels must be hydrogen-based by 2030
ReFuelEU Aviation	Legally binding sustainable aviation fuel mandate for aircraft departing EU airports	- 2% SAF by 2025, rising to 70% by 2050 - E-kerosene sub-target: 1.2% in 2030, rising to 35% in 2050.
FuelEU Maritime Directive	Legally binding Well-to-Wake greenhouse gas reduction targets for ships calling at EU ports	- 2% lower emissions compared to 2020 from 2025, rising to 80% by 2050. - Technology neutral: ammonia and methanol are options
Alternative Fuels Infrastructure Regulation (AFIR)	Legally binding requirements for construction of BEV and HFCEV fuelling infrastructure	- Hydrogen refuelling stations every 200 km along the core road network by 2030 - (EV chargers every 60 km)

pollute. Usually, a certain number of permits are awarded freely, and those who emit less can sell their surplus allowances. This creates a financial incentive to reduce emissions. The new ETS reforms expand its scope to include new sectors, reduce the total number of emissions allowances, and crucially, begin phasing out the free allowances that were previously handed out each year.

To protect European companies from unfair competition, the EU also introduced the *Carbon Border Adjustment Mechanism* (CBAM) in 2023. This requires importers of certain products to pay a fee that reflects the emissions generated during their production – effectively charging them the same carbon price that local producers must pay under the ETS. Without this, cheaper imports from countries with weaker climate rules could undercut European goods, which would undermine local industries and climate goals. The CBAM currently applies only to cement, electricity, fertilisers, aluminium, iron, steel, and hydrogen – as well as some derivative products – but it is intended to cover all imports by 2030.

Direct Investment

In addition to mandating the use of renewable hydrogen and providing guard rails for the climate integrity of its production, the EU is also providing direct investment to help get the ball rolling. Designated as Important Projects of Common European Interest (IPCEI) – a status that allows the EU to provide direct state funding for projects with strategic cross-border value – four rounds of 'Hy2' funding were awarded between 2022 and 2024. These are summarised in Table 10.3 below.

Table 10.3. Summary of hydrogen IPCEI.

Funding Round	Public Funds Available (€)	Focus Areas
Hy2Tech	5.4 billion	Research, innovation, and early industrial deployment across the entire hydrogen value chain
Hy2Use	5.2 billion	Large-scale electrolysers and hydrogen integration into industrial processes (e.g., steel, cement, glass)
Hy2Infra	6.9 billion	Hydrogen infrastructure – including electrolysers, pipelines, storage, port terminals
Hy2Move	1.4 billion	Integration of hydrogen technologies into transport modes – road, maritime, aviation

Together, the hydrogen IPCEIs have received €18.9 billion in public funding—more than half of the €37 billion allocated across all IPCEI projects up to 2025. While other areas such as batteries, microelectronics, cloud infrastructure, and medical technology have also received support, hydrogen stands out as the single largest recipient, which highlights the importance that the EU has placed upon it.

Production Subsidies

Europe's third major policy lever is the production subsidy, aimed squarely at closing the cost gap with fossil fuels. This support is being delivered through two distinct but complementary subsidy schemes: the European Hydrogen Bank, and the H2Global initiative. The Hydrogen Bank is an instrument of the European Union, which also provides its funding, whereas H2Global was launched by the German government. More accurately, the H2Global Foundation was created by a group of corporations with an interest in the hydrogen economy, and shopped to various governments until it was picked up by Germany, who now provides most of its funding.

Both schemes award subsidies through a contestable tender process, whereby companies submit the lowest subsidy they are prepared to accept, and winners are selected based on who can provide the most hydrogen, at the lowest cost. The key difference between them is that so far, the Hydrogen Bank has focused on green hydrogen production within the EU, while H2Global has targeted imports of hydrogen derivatives from outside of Europe. These tenders are run by a special purpose company—the Hydrogen Intermediary Network Company. HintCo, as it is known, was established by the German Government to manage the operations of H2Global, and was subsequently selected by the EU to do the same for the Hydrogen Bank.

While both offer hydrogen producers ten years of support, they do so in slightly different ways. The Hydrogen Bank provides conventional subsidies through fixed €/kg payments. In contrast, H2Global offers 'contracts for difference' (CfD) and uses a system of double-sided auctions. The theory is that it will import hydrogen at the lowest possible price and on-sell it to European buyers at the highest possible price. If, as is expected, the sales price is lower than the purchase price, the difference will be covered by H2Global. This method will allow H2Global to offer long-term contracts to hydrogen producers, which is what they need to confidently invest in building new process equipment. At the same time, it can offer short-term supply contracts to hydrogen consumers, who are generally afraid to lock themselves into longer deals in case the price of hydrogen should drop. Thus, H2Global can theoretically meet the contradictory needs of both buyers and sellers. We say theoretically, because it has not yet sold or received a molecule of hydrogen. Neither has the Hydrogen Bank. Both schemes are still in their infancy, and the projects that have been awarded subsidies are still under development.

The Hydrogen Bank and H2Global have both completed two tender rounds as of mid 2025, with more to come (Table 10.4). It is noteworthy that the early Hydrogen Bank auctions were both heavily oversubscribed, with subsidy requests far exceeding the available budget. While this reflects strong developer interest, it may also indicate that the support currently on offer is insufficient to support a large number of projects. It is also noteworthy that the winning bids from the first round, came in well below the generous €4.5/kg H$_2$ ceiling. The winning developers were willing to accept just €0.37 to €0.48 per kilogram of hydrogen – far less than the price gap between grey and green – suggesting they see even a modest subsidy as better than none. If all six winning projects are built, they are expected to produce just under 160,000 tonnes of green hydrogen annually.

Table 10.4. European hydrogen subsidy tenders: summary of early rounds.

Subsidy Scheme	Tender Round	Budget	Supported Products	Outcome
Hydrogen Bank	1 (2024)	€800 million; €4.5/kg ceiling	Renewable hydrogen	132 bids from 17 countries; over €12 billion requested. Winning bids from Spain, Portugal, Finland, Norway (6 in total); €720 million awarded ranging from €0.37–0.48/kg
Hydrogen Bank	2 (2025)	€1.2 billion from EU + €0.8 billion from Austria, Spain, Lithuania; €4/kg ceiling	Renewable hydrogen, (€200 million for maritime use)	61 bids from 11 countries; over €4.8 billion requested. Winners yet to be announced
H2Global	Pilot (2024)	€900 million	Imports of 1. ammonia 2. methanol 3. SAF	1. One winner; €300 million for 400 kt over 5 years 2. decision delayed, TBD 3. no bids, no awards
H2Global	2 (2025)	€2.5 billion; - Regional Lots: Africa, Asia, Nth America, Sth America and Oceania - Global Lot	- Regional bids may supply hydrogen, methanol, ammonia; - Global bids must supply hydrogen	Winners yet to be announced

The results of the early Hydrogen Bank auctions also highlighted significant differences in hydrogen production costs across Europe. According to EU's energy regulatory agency ACER, the levelized cost of hydrogen (LCOH), as averaged by country, varied from €5.3/kg in Greece to €13.5/kg in Poland. Hydro-rich Sweden and the sun-drenched plains of Spain also offer hydrogen below €6/kg, and were the 2nd and 3rd cheapest countries respectively behind Greece. With an average of €11.39/kg, Germany – the largest consumer of hydrogen in Europe – is at the pricier end of the scale.

It should be noted here that while Team Europe's posture overall is that of a buyer, the role of individual nation states does vary. For example, while Germany is a very keen buyer, countries like Spain, Portugal and the Nordic nations are more focused on exporting to their neighbours. France could arguably be placed on the list of DIY nations, with a focus that is largely on domestic production and use within industry.

In any case, the variation in LCOH by geography highlights why the cost of hydrogen transport, and consequently the buildout of pipeline networks, is so important. Centrally located industrial heavyweights like Germany and Poland will be heavily dependent on imports from the Iberian Peninsula and the Nordic countries, which are located on the edges of the continent.

International Engagement

Anticipating that it will remain partially dependant on energy imports, Team Europe has also been busily promoting renewable hydrogen production around the globe. Europe's politicians have been eagerly signing memorandums of understanding with potential hydrogen exporters, and in many cases offering funds to help get things moving. The pot of money most commonly used is the gargantuan Global Gateway fund. With €300 billion to invest between 2021 and 2027, Global Gateway accounts for 16% of the EU budget. Its broad purpose is to contribute investment to developing countries, which generally have a harder time attracting it than rich countries, while also creating opportunities for European companies to invest and grow their wealth. Approximately half of the €300 billion has been dedicated to Africa. The other regions of focus are Asia Pacific; and Latin America and the Caribbean.

The Global Gateway is not exclusively focused on hydrogen of course. In addition to 'Climate and Energy', the fund also targets investment in education and research; the digital sector; transport; and health. However, hydrogen has definitely been receiving a piece of the pie. The largest recipient so far appears to be Brazil, which has been promised €2 billion to produce green hydrogen and promote energy efficiency in industry. A combination of grants and loans to the tune of a few hundred million each have also been offered to its neighbours – Argentina and Chile. Meanwhile in Africa, Namibia has received significant attention with the EU promising to mobilise €1 billion of investment for renewable hydrogen and raw material infrastructure, which includes a €120 million grant for hydrogen. Other African nations to receive offers of hydrogen investment dollars, either by the EU or individual European nations, include Algeria, Kenya, Morocco, Mauritania, and South Africa.

While not related to the Global Gateway, Team Europe has also made a substantial investment in India. The European Investment Bank (EIB) has pledged to provide funding of up to €1 billion for large-scale green hydrogen projects. The EIB is the EU's lending arm, providing financing for projects across Europe and globally, focusing on economic development, climate action, and infrastructure.

Policy Criticisms

In terms of hydrogen policy initiatives, Europe has it all. Subsidised production, subsidised imports, consumption mandates, and measures to protect domestic industry are all on the card. While this portfolio is unrivalled in scale and ambition, it is far from certain that its goals are achievable, affordable, or even necessary.

Firstly, it is questionable whether the 2030 targets of producing 10 Mt and importing another 10 Mt are justified. These are clearly 'policy numbers' – selected because they are memorable and politically desirable. Europe's current hydrogen consumption is only around 7 Mtpa, barely a third of the targeted 20 Mtpa. The first hydrogen monitoring report from ACER warns that the scale of Europe's planed continent-wide hydrogen pipeline network is based more on political hopes than actual demand – raising the risk that parts of it could end up underused or stranded.

Similarly, it has been calculated that the renewable hydrogen required to meet Europe's legally binding mandates in 2030 amounts to just 3.5 Mt for industry and 1 Mt for transport – less than a quarter of the implied 20 Mt consumption target.[2] What's more, Europe should be capable of meeting these binding mandates through domestic production. The engineering consultancy Ricardo also estimates that the continent could produce between 6 and 7.5 Mt of renewable hydrogen by 2030, which would avoid the need for expensive imports.

However, even achieving this lower level of production appears unlikely. ACER concluded that the bloc is well off track in scaling up renewable hydrogen. Producing 6–7.5 Mt would require a massive expansion of renewable electricity generation and electrolyser capacity. For instance, 7.5 Mtpa would consume nearly 60% of Europe's current renewable electricity output and require approximately 75 GW of electrolysers.[3] Europe's installed electrolyzer capacity is a mere 0.2 GW, with a further 1.8 GW under construction. ACER notes that proposed projects could add an additional 60 GW by 2030. However even if all of these were completed, which is highly unlikely based on recent failure rates, Europe would still fall short by 20%. Industry body Hydrogen Europe has reported that only 21% of projects that had received funding under the Hy2 IPCEI scheme had reached final investment decision as of early 2025.[4]

This lack of progress has triggered political pushback against the EU's climate guardrails for hydrogen. In late 2024, Germany's then Vice-Chancellor and Minister for Economic Affairs and Climate Action, Robert Habeck—one of Europe's strongest hydrogen champions—formally asked the European Commission to delay implementation of the additionality and time-matching rules. While noting that he had supported the framework, he wrote that "reality has now shown that these requirements were too high and are slowing down the ramp-up of the projects for the production of renewable hydrogen." About six months later, *Hydrogen*

Insight reported that a coalition of twelve EU member states had also called for a reassessment of the rules, questioning whether they were fit for purpose. Clearly, lawmakers are beginning to second-guess whether the original protections were too strict, given that the hoped-for hydrogen buildout has been slow to materialize.

The intention was admirable: to ensure climate integrity while reducing price. However making green hydrogen in Europe remains costly. According to ACER, the cost of "renewable hydrogen" that meets the RFNBO definition averages around €8/kg across Europe. This figure aligns with commercial reports from 2024, which suggested that buyers seeking long-term renewable hydrogen supply were typically seeing offers around €8/kg – well above the €1/kg cost of grey hydrogen.

Providing enough subsidy money to close this price gap presents formidable financial and political hurdles. Based on early Hydrogen Bank tender results – where winning bids ranged from €0.37–0.48/kg H_2 – scaling domestic hydrogen production to 10 Mtpa would require €4–5 billion in subsidies per year. As a point of comparison, EU institutions contributed nearly €27 billion to international development aid in 2023 according to Donortracker.org. Both are investments aimed at addressing global challenges, and a hydrogen subsidy of this level appears to be a manageable addition to the EU's portfolio of public-good spending.

On the other hand, H2Global's award of €300 million to support the delivery of 400,000 tonnes of ammonia – from a facility in Egypt to the Dutch port of Rotterdam over five years – implies a subsidy of €4.3/kg H_2. At this level, which is close to the Hydrogen Bank's ceiling, 10 Mtpa of imports would require an annual budget in excess of €40 billion. This much larger figure is unlikely to be politically palatable. The disparity between subsidy levels makes clear that unless the cost of delivering clean hydrogen falls dramatically, Europe's most ambitious targets are unlikely to be affordable.

Despite its unrealistic targets, the EU's mandates position it as arguably the biggest driver of clean hydrogen worldwide, and this looks likely to remain the case for some time. The President of the European Commission, Ursula von der Leyen, is a strong supporter of hydrogen and was re-elected at the end of 2024 for a second five-year term. In her words, "Rome was not built in a day, and neither will a world-leading, climate-neutral European economy. But together, step by step, we are building it. And the European Commission is fully committed to staying the course."

Japan and The Republic of Korea

Japan and Korea were among the first to publish national hydrogen strategies, and have since placed hydrogen and its derivatives at the core of their plans to reach carbon neutrality by 2050. Their reasons for doing so are also very similar. Firstly, the electricity sector is the largest source of greenhouse gas in both countries. The IEA reports that in 2022, the power and heat sector was responsible for 48% of Japan's emissions and 52% of Korea's. Decarbonising power generation is an important goal for both nations.

They also share a perceived challenge in meeting their electricity needs through renewable energy. This is due to both nations having dense populations; a relatively small land area; and a geographic location that precludes importing electricity directly from trustworthy neighbours. Indeed, their locations are nearly identical. Korea's port of Ulsan, which received the world's first shipment of blue ammonia, and Japan's port at Kobe, which received the world's first shipment of liquid hydrogen, are only about 500 km apart as the crow flies.

While it is true that neither Japan nor Korea has an abundance of available land, the claim that they cannot produce enough renewable energy is questionable, given they both have substantial offshore wind resources. Recent studies suggest that Japan can theoretically meet all its electricity needs with domestic renewables and energy storage, and that Korea could reach at least 80% by 2035.[5, 6] Despite this, the governments of both countries are pushing on under the assumptions that a massive build out of renewables will be too hard or take too long, and that importing clean energy in the form of hydrogen will be necessary to meet their net zero targets.

The hydrogen policies of Japan and Korea are slightly unusual in the global context because rather than focus on incentivising production, as is the case in most places, they are attempting to establish domestic hydrogen use cases under the belief that they will be able to meet this demand through imports. Energy independence is therefore not on their list of reasons for developing a hydrogen market. Rather, they are hoping to achieve energy security by having a diverse network of hydrogen suppliers.

To this end, both nations have been actively seeking partnerships around the globe. Various Japanese government agencies and corporations have signed hydrogen-related agreements with partners in Europe, the Pacific Rim (Australia, Chile, Indonesia, and New Zealand), and the Middle East (Saudi Arabia and the UAE). Korea also has hydrogen-related agreements with Australia and Saudi Arabia, as well as a technology cooperation pact with the United States. Naturally enough, Japan and Korea are working together as well. In 2024, they held their first bilateral hydrogen dialogue, which established a framework for cooperation on supply chains, investment, and the development of hydrogen technologies and standards.

Some concerns have been raised about whether enough specialised ships could be built in time to support their import plans. The International Chamber of Shipping has estimated that Japan and South Korea's combined 2030 targets would require around 70 hydrogen carriers—none of which currently exist beyond the pilot-scale *Suiso Frontier*.[7] However, most plans for transporting clean energy rely on ammonia rather than liquefied hydrogen, and industry reports suggest that almost 90 new ammonia-capable vessels are set to join the global fleet by 2028. While other countries will no doubt compete for deliveries, this suggests that shipping capacity should more or less align with Japan and Korea's near-term needs – as long as they can use ammonia rather than hydrogen.

With that context in mind, we can now turn to the specific strategies being pursued by each country, beginning with Japan.

Japan

Japan's hydrogen strategy relies heavily on international partnerships. It has invested in projects such as the Hydrogen Energy Supply Chain (HESC) in Australia, which aims to produce hydrogen from brown coal with carbon capture, liquefy it, and ship it to Japan. Similar initiatives are being explored with countries such as Brunei, Saudi Arabia, and the UAE. These efforts reflect Japan's longstanding strategy of diversifying its energy imports rather than aiming for full energy independence.

As mentioned previously, Japan's Basic Hydrogen Strategy, released in 2017, was the first national hydrogen strategy in the world. It reflected not only environmental ambition but deep concerns about energy security in a relatively resource-poor nation still reeling from the 2011 Fukushima disaster. Hydrogen, and its derivatives such as ammonia, are seen as a way to diversify energy imports and maintain industrial competitiveness in a decarbonizing world.

Japan's 2021 Green Growth Strategy for Achieving Carbon Neutrality by 2050 further highlights this focus, listing "hydrogen and fuel ammonia" as one of 14 critical sectors for investment. Indeed, both strategies tend to use "hydrogen" as shorthand for "hydrogen and its derivatives". Together, these two strategies define key targets for the coming decades.

Japan currently generates about a third of its electricity from coal and another third from natural gas. Renewables account for just under a quarter of the mix, while nuclear provides around 8%. Looking ahead, the government aims to shift the balance substantially. By 2050, it hopes to generate 50–60% of electricity from renewables, with a further 10% coming from hydrogen.

To put this into perspective, Japan currently consumes around 1,000 TWh of electricity per year. Meeting 10% of that demand with hydrogen would require roughly 6 Mtpa, assuming it is converted into electricity via a fuel cell or turbine operating at around 50% efficiency. This level of hydrogen consumption would account for about a third of Japan's projected hydrogen supply in 2050, based on its current targets.

Japan is flexible about where this hydrogen comes from. It may be produced domestically or imported, with the government aiming for Japanese companies to control 15 GW of electrolyser capacity by 2030—about 4–9% of expected global capacity, according to IEA estimates. However, there is little insistence that the hydrogen must be electrolytic: Japan remains open to supply from fossil fuel-based production with carbon capture.

Substantial direct investment is being funnelled towards engineering solutions to the challenges standing in the way of an international hydrogen trade. Japan's Green Innovation Fund, endowed with a ¥2 trillion ($16 billion) budget over a decade, has earmarked roughly 40% of the money for hydrogen-related projects, including:

- $560 million to advance electrolysis technology
- $2.4 billion for large-scale hydrogen transport, liquefaction, and combustion demonstrations
- $3.5 billion for hydrogen use in iron and steel production, particularly through direct reduction

Additional investments include up to $550 million for an ammonia combustion supply chain, $230 million for hydrogen-fuelled ships, and an eye opening $34 billion dedicated to hydrogen-based aviation in addition to $110 million for the development of hydrogen fuel cell propulsion systems. It hopes to achieve domestic production of a "next generation" regional aircraft by 2035.

Japan is also supporting early demonstrations of hydrogen-fired gas turbines, with an eye on future export markets, and research into stationary fuel cells.

Altogether, the government aims to mobilize ¥15 trillion ($100 billion) in public and private investment over the next 15 years. It hopes this will drive down hydrogen costs to a level where it is competitive with fossil alternatives – about $2.20 per kilogram by 2030, falling further to $1.45 per kilogram by 2050.

Several major Japanese companies are deeply involved in these efforts. Kawasaki Heavy Industries is leading development of hydrogen shipping and liquefaction technologies. JERA, Japan's largest power generator, is piloting ammonia co-firing at coal plants. Toyota and Honda continue to back hydrogen fuel cell vehicles despite stagnant global sales.

However, some of Japan's technology bets remain speculative and controversial. Hydrogen-based aviation faces daunting technical and economic hurdles, as discussed in Chapter 7. Ammonia co-firing in coal plants has been criticized for risking further entrenchment of fossil fuel infrastructure, and Japan's promotion of these technologies in Southeast Asia has drawn accusations of undermining global decarbonization efforts, as discussed in Chapter 6.

Beyond direct investments, Japan has also announced its intention to enter the subsidy game. In May 2024, it introduced the Hydrogen Society Promotion Act, allocating ¥3 trillion ($19.25 billion) to a Contracts for Difference program. Under this scheme, the government will pay clean hydrogen producers the difference between the market price and a government-set reference price. Contracts will last for 15 years—significantly longer than similar programs elsewhere, which typically top out at 10 years.

Notably, companies producing hydrogen overseas may be able to stack Japanese subsidies on top of domestic ones, allowing for "double dipping" into multiple government supports. This mechanism could be powerful in driving down consumer prices and positioning Japan, alongside the EU, as one of the principal architects of a theoretical international clean hydrogen market.

South Korea

With the 2013 release of its Tucson ix35, Hyundai claimed to have produced the world's first mass-market hydrogen fuel cell electric vehicle. By 2015, POSCO Energy, a subsidiary of Korea's largest steelmaker, had built the world's largest fuel cell manufacturing plant. Against this backdrop of early fuel cell leadership, President Moon Jae-in – newly elected in 2018 – pledged to turn the Republic into a "hydrogen economy".

Korea's 2019 Hydrogen Economy Roadmap was quickly followed by the world's first hydrogen-specific law in 2020. The *Hydrogen Economy Promotion and Hydrogen Safety Management Act* created a legal framework for government support, including subsidies and standards. However, unlike other nations primarily motivated by decarbonization and energy security, Korea's plan was focused on economic growth and cementing its industrial lead in fuel cell technology.

This was evident in the roadmap's targets. By 2040, Korea aimed to produce 15 GW of hydrogen fuel cells for power generation, with roughly half intended for export. It also planned to build 1,200 hydrogen refuelling stations and put 6.2 million FCEVs on the road, again with over half targeted for foreign markets. Generous subsidies from both national and city governments halved the cost of FCEVs and refuelling infrastructure, propelling Korea to world leadership in hydrogen cars on the road.

The cleanliness of hydrogen was not a primary concern at the time. In fact, the roadmap implied continued reliance on grey hydrogen—which remains Korea's dominant source today. As a result, Korea's FCEV fleet is only delivering around 30% CO_2 savings compared to conventional petrol cars. Recognizing that a grey hydrogen economy was inconsistent with its climate obligations—and that an exclusive focus on fuel cells had left it lagging in other aspects of hydrogen technology, Korea revised its policy in 2022 under the new Yoon administration.

The updated '3-Up' hydrogen strategy centred on three pillars:

Scale-Up. Expanding the clean hydrogen ecosystem by building a global supply chain and creating large-scale demand, particularly in power generation and transportation.

Build-Up. Constructing infrastructure to support the hydrogen economy, including the world's largest liquid hydrogen plant, 70 liquid hydrogen fuelling stations, new port facilities for ammonia and hydrogen imports, and designated 'hydrogen cities' to demonstrate integration into daily life.

Level-Up. Investing in innovation with the aim of becoming a global leader in hydrogen technology, including offering tax credits for hydrogen research and development.

Given Korea's position as a manufacturing powerhouse, it is notable that the 3-Up strategy for creating hydrogen demand is focused on controversial use cases—transportation and power generation—rather than sectors that already use hydrogen or are difficult to electrify.

In addition to the original 2040 target of 6.2 million FCEVs, Korea has also set a nearer-term goal of 300,000 by 2030, 7% of which should be buses. As of 2024, the country had around 35,000 FCEVs on the road and about 1,000 buses—numbers that make the targets look ambitious to the point of improbability, especially given collapsing sales in recent years, as discussed in Chapter 7.

Recent safety issues have only compounded the problem. In 2024, a global recall of the Hyundai Nexo for potential hydrogen leaks coincided with revelations that 11% of Korea's hydrogen buses were also leaking—a discovery made after a hydrogen refuelling station caught fire and a bus exploded, injuring three people, in the same week.

It has not been smooth sailing for Korea's power generation ambitions either. Like Japan, Korea currently generates roughly one-third of its electricity from coal and another third from natural gas. However, it is further behind on renewables (9%) and relies more heavily on nuclear power (29%). The government has targeted 2.4% of national electricity to come from clean hydrogen and ammonia by 2030, increasing to 5.5% by 2038. To help achieve this, it introduced the Clean Hydrogen Portfolio Standard (CHPS), under which the government offers 15-year power purchase contracts to generators supplying hydrogen-based electricity.

Power companies are permitted to use fuel cells; co-fire up to 50% hydrogen in natural gas turbines; or blend up to 20% ammonia into coal-fired power stations. Contracts are awarded via tender based on price and emissions reduction potential. Two pilot tenders in 2023 awarded a modest 1,300 GWh of contracts—less than a day's electricity consumption in South Korea. Although results were not officially disclosed, reports suggest that many of the winning bids involved fuel cells powered by grey hydrogen. Assuming grey hydrogen emissions of 10 t CO_2/t H_2, and a fuel cell efficiency of 55%, this results in a carbon intensity of around 500 t CO_2/GWh—similar to, or even slightly worse than, burning natural gas in a turbine.

In 2024, Korea ran the world's first auction for "clean hydrogen power." This time, 6,500 GWh of contracts were on offer, with grey hydrogen use excluded. The result was dismal. Only one contract was awarded: Korea Southern Power Company is to supply 750 GWh per year by co-firing ammonia in a coal plant. This is just 12% of the energy that the government had hoped to procure. Reporting by *S&P Global Commodity Insights* suggests that the high cost of clean hydrogen rendered most bids unaffordable.[8] Although the government did not disclose how much it was prepared to pay, the industry has estimated that the figure was around $0.35/kWh, which is four times the usual wholesale electricity price. This implies an ammonia price around $640/tonne. However, the bidding power companies could only acquire ammonia at $700–800/tonne. In short, even allowing for quadrupled fuel prices, clean hydrogen-derived power is currently not viable.

Coincidentally, around the same time the auction results were unveiled, and hydrogen buses were bursting into flames, President Yoon found himself engulfed in a conflagration of his own, in an episode widely covered by the international press. In December 2024, Korea was thrown into chaos when the president attempted to declare martial law, citing vague threats to national security. As members of the National Assembly rushed to convene an emergency session, they were physically blocked by security forces, leading to a tense standoff outside the legislative building. Ultimately, the defiant lawmakers managed to force their way past police barricades and convened an emergency vote to reject the decree and maintain their democracy. Under immense pressure, Yoon rescinded martial law within hours, but the damage was done. Days later, the National Assembly overwhelmingly voted to impeach him, stripping him of power and setting off a historic legal battle over his fate. In the wake of these dramatic events—and mounting scepticism over Korea's hydrogen strategy—some analysts expect the next government to take a different approach.

A Costly Strategy?

The import-heavy strategies of our buyer nations are based on an assumption that clean hydrogen, along with the infrastructure needed to transport it, will be widely available at an affordable price. This is a brave assumption to make—and a risky strategy to build upon, given the realities of hydrogen production and logistics, which have already been detailed in these pages.

To illustrate just how expensive imported hydrogen energy might be, we draw on 2030 cost projections developed by the U.S. Department of Energy (DOE), and illustrated in Figure 10.1. These estimates are particularly useful because they include not only the cost of producing hydrogen—which depends heavily on the quality of renewable resources—but also the cost of building, operating, and maintaining the infrastructure required to move hydrogen between continents. Importantly, the DOE's analysis is focused on Japan, South Korea, and Europe, providing estimates of delivered hydrogen prices from a range of likely exporters.

It is worth noting, however, that the DOE's conclusion—that U.S. producers could supply hydrogen at the lowest cost—is not universally shared. For example, the IEA's analysis suggests that regions such as the Middle East, Latin America, and parts of China, among others, may achieve lower production costs. From the perspective of the buyer, however, the origin of the hydrogen is less important than its final price. On this basis, the DOE's work remains valuable, showing that even under optimistic assumptions, importing nations will probably face costs in the range of $4–7/kg, just to bring hydrogen into their ports.

To put these estimates into context, they must be compared to the cost of conventional energy carriers. Table 10.5 provides such a comparison for electricity and natural gas. While energy prices fluctuate by season and vary between countries, these figures offer a reasonable average for Europe, Japan, and South Korea.

The table shows that if imported hydrogen were used directly to produce heat for an industrial process, it would at minimum double the user's energy costs, and could more than triple them if import prices end up at the higher end of the projected range. The same holds true for electricity. Assuming hydrogen is converted to electricity via a gas turbine or stationary fuel cell at around 50% efficiency, the implied cost of hydrogen-fuelled electricity would be three to four times higher than that of conventional generation.

To be fair, no one is proposing to generate all electricity from hydrogen, so the impact on consumer prices depends on the share hydrogen holds in the energy mix. If hydrogen were to supply 10% of electricity—as envisioned in Japan's 2050 target—the overall electricity price would increase by 20–30%. That is not as catastrophic as it might first appear, but neither is it trivial.

These comparisons highlight a critical point: importing hydrogen is not simply a technical exercise in substituting one fuel for another. It entails accepting a significant and lasting increase in energy costs across key sectors.

Table 10.6 presents a similar analysis for the case where hydrogen is imported as a vehicle fuel. As with electricity and gas, the price of petrol in Japan and South Korea is broadly similar, while the range across European countries is wider and generally somewhat higher, largely due to taxes.

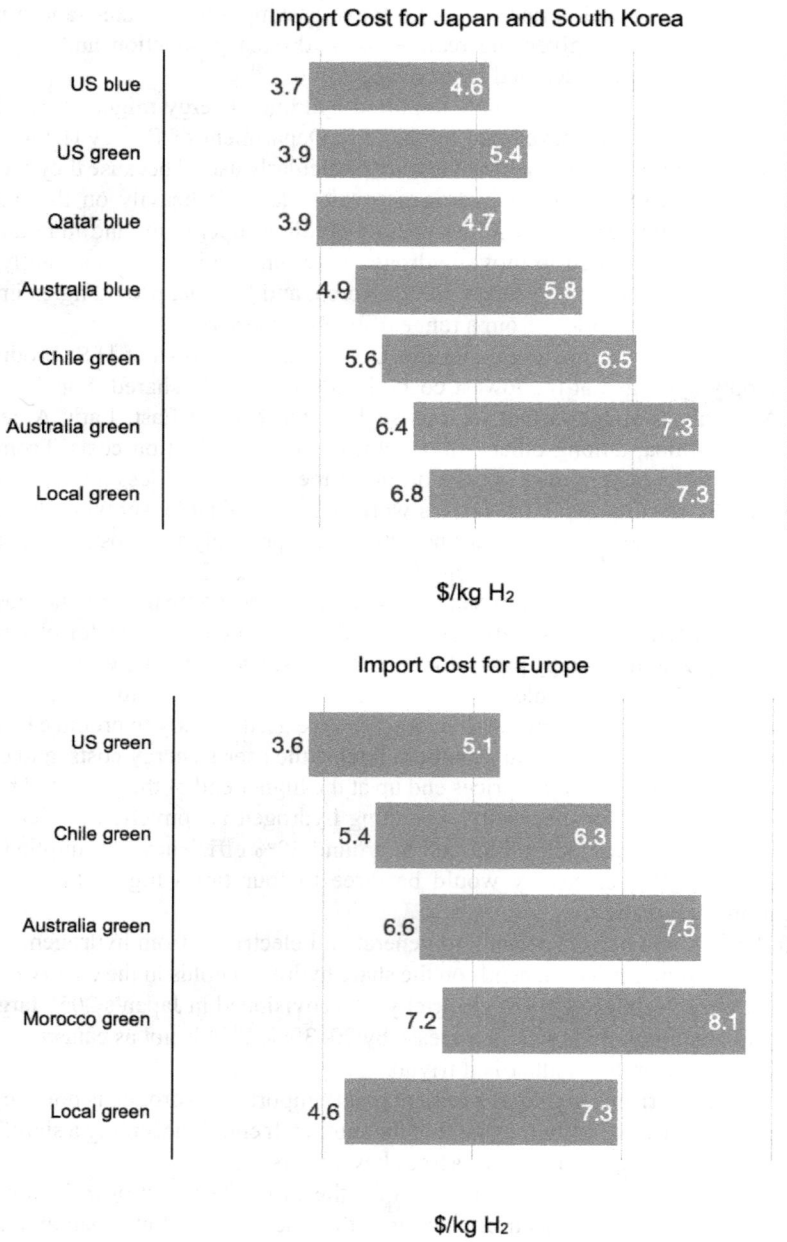

Figure 10.1. 2030 Cost estimates for imported clean hydrogen for Japan, South Korea, and Europe. Source: US DOE (2024).[9]

Table 10.5. Projected 2030 energy costs for Japan, South Korea, and Europe.

	Imported Hydrogen[1]	Electricity: Conventional Generation[2]	Electricity: Hydrogen Generation	Heat: Natural Gas[3]	Heat: Hydrogen Gas
	$/kg	$/kWh	$/kWh	$/kWh	$/kWh
Low Estimate	4	0.09	0.24 (3x higher)	0.06	0.12 (2x higher)
High Estimate	7		0.42 (4x higher)		0.21 (3.5x higher)

Source: 1. US DOE (2024).[9] 2. IEA (2024).[10] 3. IEA (2025).[11] Note: 2 and 3 adjusted for 2% annual inflation to 2030.

Table 10.6. Projected 2030 fuel costs for road transport in Japan, South Korea, and Europe.

	Petrol Price[1]	Mobility Cost: ICE Vehicles[2]	Hydrogen Price[3]	Mobility Cost: FCEV[2]	Hydrogen Premium
	$/L	$/100 km	$/kg	$/100 km	
Japan	1.4	7.7	8–14	8–14	1–2x
South Korea	1.5	8.3			1–2x
European Range	1.7–2.3	9.4–12.7			1.6–0.9x, 1–1.5x

Source: 1. Globalpetrolprices.com (April 2025). Notes: 2. Assumes efficiency of 5.5 L/100 km of ICE; 1 kg/100 km for FCEV. 3. Assumes hydrogen retail price is twice the import cost.

Assuming a hydrogen pump price of $8–14 per kilogram—which is actually lower than retail prices seen in recent years—the premium that consumers would pay for hydrogen mobility is less severe than in the case of electricity or heating. Thanks to the higher efficiency of fuel cells compared to combustion engines, hydrogen could approach price competitiveness with petrol in European countries with the highest fuel taxes. In other regions, however, hydrogen would still likely be up to twice the price of petrol.

Of course, as discussed in Chapter 7, all the fundamental reasons why battery electric vehicles outperform fuel cell vehicles still apply. Even if hydrogen fuel were to achieve parity with petrol, it would remain second-best as a decarbonisation option. Across electricity, heating, and transport, the story is the same: imported hydrogen offers little hope of delivering cheap, competitive energy in the near term. Government subsidies can reduce the price paid by end users, but the full cost must still be covered—either through taxes or public debt.

So, while Europe, Japan, and South Korea are preparing to pay a premium for clean hydrogen, a growing number of countries and companies are positioning themselves to supply it. For these prospective sellers, the hydrogen economy is not a cost to be managed, but an opportunity to be seized. We turn now to the sellers.

Big Numbers: Chapter 10

- **25 million tonnes** - Combined 2030 consumption targets of the European Union, Japan, and South Korea - the driving force behind clean hydrogen trade.
- **$7** - Average price of green hydrogen in Europe as of 2025. Grey hydrogen is $1–2.
- **$100 billion** - Total investment Japan hopes to mobilize by 2040.
- **6.2 million** - Korea's 2040 fuel cell electric vehicle target. Hydrogen strategy has a strong focus on industrial development.
- **70** - Number of hydrogen carrier vessels needed to meet Korea and Japan's 2030 import targets. None exist as of 2025.
- **$4–7** - Projected cost per kilogram of clean hydrogen imports in 2030. Expensive energy even at a reduced price.

CHAPTER 11

The Sellers

⬦⬦

Hydrogen's would-be exporters are a diverse group of countries hoping to leverage their natural resources and energy expertise to supply clean energy to the world. Some, like Australia, Canada, and the Gulf states, are already major players in fossil fuel markets and see hydrogen as an opportunity to sustain their export economies in a decarbonising world. Others—notably in Africa and Latin America—are developing economies who see hydrogen as a pathway to industrial development and foreign investment.

Across this broad group, there are some common themes. All are rich in renewable energy potential, all see hydrogen as a vehicle for economic opportunity, and all face the same fundamental challenge: turning aspirational strategies into real projects. These nations have set ambitious targets for 2030 and beyond and have attracted significant project proposals. However, almost none of these proposals have progressed to a final investment decision. As of 2025, international hydrogen trade remains largely theoretical. We'll now touch on the strategies, incentives, and targets each of them will employ, beginning with the higher-income nations as summarized in Table 11.1.

High-Income Nations

Table 11.1. Hydrogen strategy summary: high income export-focused nations.

	Strategy Document	Key Policy Instruments	Notable Targets
Australia	November 2019; National Hydrogen Strategy	10 year hydrogen production tax credit of $1.30/kg H_2	By 2030: 0.5–1.5 Mtpa made 0.2–1.2 Mtpa exported By 2050: 15–30 Mtpa made
Canada	December 2020; Hydrogen Strategy	Clean hydrogen production tax incentive on plant costs until 2034: 40% if < 0.45 $kgCO_2/kgH_2$ 25% if 0.75–2 $kgCO_2/kgH_2$ 15% if 2–4 $kgCO_2/kgH_2$	By 2050: Top 3 hydrogen producing nation, H_2 replaces 50% of natural gas consumption
Oman	October 2022; Green Hydrogen Strategy	50,000 km² of land rights auctioned	Produce: 1.3 Mtpa by 2030 3.75 Mtpa by 2040 8.5 Mtpa by 2050
Saudi Arabia	No formal strategy	State-owned developers	Produce: 2.9 Mtpa by 2030 4 Mtpa by 2035
United Arab Emirates	November 2021; Hydrogen Leadership Roadmap	State-owned developers	2031 production of 1.4 Mtpa 'low carbon' hydrogen

DOI: 10.1201/9781003361428-14

Australia

Australia is one of the world's largest energy exporters. In 2023, it led the world in coal exports and ranked fifth in natural gas, supplying mostly to Asia. Blessed with abundant land for large-scale projects and world-class solar and wind resources, Australia is also a theoretically ideal location for the build-up of renewable energy production. However, with 47% of its electricity coming from coal in 2023, along with another 20% from oil and natural gas, Australia is still a long way from decarbonising its own grid. Nevertheless, the hydrogen-buying nations came knocking early.

Even before Australia launched a formal hydrogen strategy, it had partnered with Japan on the Hydrogen Energy Supply Chain Project (described in Chapter 7). Recognizing the momentum building around clean hydrogen, the Morrison government released Australia's first National Hydrogen Strategy in 2019, aiming to establish hydrogen as its next major export.

By the end of 2021, Australia had inked agreements with most of the world's would-be hydrogen buyers. It signed a Joint Statement of Cooperation on Hydrogen and Fuel Cells with Japan, committed to advancing supply to Europe through the Australia-Germany Hydrogen Accord, and entered into partnership with Singapore to advance low-emissions fuels for maritime and port operations. Additionally, it established a Zero Emissions Technology Partnership with Korea, pledging joint efforts in developing and commercialising hydrogen technologies, including fuel cell vehicles and hydrogen power generation.

Enthusiasm for clean hydrogen only increased when the Albanese Labour government took over in 2022. Frustrated that only 10 MW of Australian green hydrogen production had reached final investment decision, compared to 1,378 MW in the EU and 280 MW in the US, the new government introduced the Hydrogen HeadStart initiative. Initially allocated $1.25 billion, and later expanded to $2.5 billion, this program aims to support large-scale green hydrogen projects by contributing to their operational costs. It is hoped that this will bridge the price gap between renewable hydrogen and fossil fuels for two or three flagship projects by 2030.

Support for hydrogen was further strengthened with an updated 2024 National Hydrogen Strategy. The 2030 goal is now set at 0.5–1.5 million tonnes produced, with 0.2–1.2 million tonnes exported. A subtle shift in focus also suggests the government has decided to put more weight on domestic hydrogen consumption, due to the difficulty of getting purely export-focused projects built. It identifies green metals (particularly iron and alumina), ammonia, long-haul transport, and power generation as local demand sectors that could help the industry scale up.

This new strategy was backed by additional funding through the Future Made in Australia package, which committed $5 billion over the next decade to boost hydrogen production, technology development, and workforce training. The Hydrogen Production Tax Incentive accounted for $4.2 billion of this, offering

$1.30 per kilogram of hydrogen produced for ten years. However, the independent Australia Institute has questioned how impactful this will be, estimating that the subsidy would support only 550,000 tonnes of hydrogen production per year, which is barely more than Australia's existing hydrogen consumption and would likely be absorbed by existing users, leaving little room for industry expansion or new export capacity.[1]

It is fair to say that despite consistent government support since 2018, Australia has struggled to get major projects over the line, and headline-grabbing cancellations have increased uncertainty around the future of clean hydrogen. In October 2024, Origin Energy, one of Australia's 'big three' energy companies, announced that it was withdrawing from all hydrogen development opportunities, citing uncertainty about market conditions. Not long after, the Queensland Government withdrew its support for the $7.5 billion flagship QC-H2 project, again over economic concerns. This withdrawal effectively cancelled the project, which was supposed to be Australia's largest green hydrogen facility on completion. Further, recent analysis by consultancy Rystad Energy, as reported by *Hydrogen Insight,* indicates that over a third of all planned green hydrogen capacity in Australia carries a high risk of not reaching completion, underlining the fragility of the sector.

Canada

Like Australia, Canada is a major exporter of fossil fuels. In 2023, it was fifth on the global leaderboard for sales of both coal and oil. Unlike Australia, however, Canada generates most of its domestic electricity from clean sources. In 2023, two-thirds of its power came from renewables, primarily hydroelectricity. A low-carbon baseload supply is ideal for electrolytic hydrogen production, allowing plants to operate continuously with minimal associated emissions. This places Canada in a strong position to produce hydrogen products that meet the highest environmental standards.

The Hydrogen Strategy for Canada was released in December 2020 after three years of consultation. Its stated purpose was to "leverage the momentum [of global hydrogen enthusiasm], to grow the domestic opportunity for hydrogen, while also benefiting from growth in global demand through export opportunities." It offered a broad list of domestic use cases without clear prioritization. The vision for 2050 includes Canada as one of the top three hydrogen-producing nations in the world. By this time, hydrogen should have replaced 50% of Canada's natural gas consumption, and make up 30% of its overall energy supply. It also targets a hydrogen price range of $1.00–$2.50 per kilogram.

The Canadian government's main vehicle to drive its hydrogen ambition is the Clean Hydrogen Investment Tax Credit (CH ITC). This provides companies building hydrogen plants a tax credit on equipment and construction costs, ranging from 15–40%. The exact amount of the credit depends on the life cycle emissions of the hydrogen produced. Hydrogen with less than 0.75 kg CO_2/kg H_2 may receive the full 40%, while anything above 4 kg CO_2/kg H_2 will be ineligible for support. This credit is available to all low-carbon hydrogen production methods, including blue hydrogen from natural gas with carbon capture and pink hydrogen from nuclear power. This scheme will run through 2035, and the government anticipates that it will cost around \$12.5 billion in total.

Canada has also entered into a bilateral agreement with Germany to develop a long-term clean hydrogen supply chain between the two countries. In August 2022, this Canada-Germany Hydrogen Alliance committed to launching aligned supply and demand auctions, through which hydrogen purchase agreements between Canadian producers and European buyers will be secured. The German-backed H2Global Foundation will run the auctions as part of its program, with a \$200 million contribution from the Canadian government to help subsidize purchases.

It is hoped that the high-quality wind resources around Canada's Atlantic coast, combined with the relatively short shipping distance to Europe, will provide the latter with a valuable source of clean hydrogen. According to Canada's 2024 Hydrogen Strategy Progress Report, the country had around 80 low-carbon hydrogen projects at various stages of consideration or development. According to its developer Everwind, the likely first recipient of Canada's CH ITC will be the Point Tupper project in Nova Scotia. It aims to begin production in 2026 and eventually export up to 500,000 tonnes of green ammonia per year to Europe.

Canada has many of the right ingredients for a thriving hydrogen industry—abundant clean power, policy incentives, and access to major markets. Yet turning these advantages into large-scale production remains a work in progress.

The Middle East

It's no secret that the wealth of the Middle East has been built on exporting its massive stock of fossil fuel resources. Cognizant of the need to diversify their economies in a decarbonising world, some nations are looking at clean hydrogen as a partial solution. The region's abundant solar and wind resources, along with existing expertise in large-scale energy exports, make it well placed in this regard.

According to IEA analysis, Oman, the United Arab Emirates (UAE), and Saudi Arabia are expected to lead hydrogen exports from the Middle East. Oman in particular stands out, potentially accounting for 60% of the region's hydrogen exports by 2030, with the UAE and Saudi Arabia projected to contribute around 20% each.[2] Unlike the subsidy-heavy approaches seen in many other nations, government support in the region has been more subtle, partly because many of its energy companies are state-owned, making it easier to align action with policy.

The UAE's strategy targets two "hydrogen oases" by 2031, which should produce one million tonnes of green hydrogen and 400,000 tonnes of blue. It forecasts that local demand will reach 2.7 million tonnes by this time, suggesting that domestic use may take priority over exports in the near term. Longer term, the UAE anticipates it will produce 7.5 Mtpa by 2040, doubling that by 2050.

The Middle East's first hydrogen-based green steel was produced at a pilot plant in Abu Dhabi in 2024. Masdar, a UAE state-owned renewables company, was a partner in the project and had pledged to develop one million tonnes of green hydrogen capacity by 2030, before pushing the target back to 2034. Oil producer Adnoc – another state-owned company – is constructing a blue ammonia facility in Abu Dhabi, which is expected to produce one million tonnes per year.

Interestingly, the region's largest energy exporter, is yet to release a formal hydrogen strategy, despite making bold claims about becoming the world's leading producer. Saudi Arabian Energy Minister Prince Abdulaziz bin Salman stated at a 2020 press briefing that Saudi Arabia "will not be challenged in its record of being the biggest exporter of hydrogen on earth."[3] Despite the lack of a hydrogen-specific strategy, the 2021 Saudi Green Initiative includes a goal of producing four million tonnes of clean hydrogen by 2030.

The country's flagship hydrogen project is the 2.2 GW green hydrogen and ammonia complex at Neom, already under construction and set to become the first gigawatt-scale green hydrogen plant in the world. Meanwhile, state-owned Aramco had announced a target of 11 million tonnes of blue ammonia production by 2030. However, it was widely reported in early 2025 that this target had been scaled down to around 2.5 million tonnes, with the shift attributed to difficulties in securing long-term buyers at a viable price.

Oman has arguably been the most eager to develop a hydrogen export industry, despite its smaller oil and gas sector. To support its production targets, which ramp up to 8.5 Mtpa by 2050, the government has auctioned around 50,000 km² of land to hydrogen developers. This area is somewhere between the size of Slovakia and Costa Rica. In 2024, the first auctions awarded land parcels to six projects, with a combined capacity exceeding one million tonnes per year, bringing Oman's 2030 target of 1.3 Mtpa within reach.

With the exception of 200,000 tonnes intended for green steel production, these projects all plan to export green ammonia. However, none have yet reached final investment decision or received full regulatory approval and it remains to be seen how many will progress from concept to reality.

Low and Middle-Income Nations

In contrast to the Gulf petro-states, many of the countries pursuing hydrogen export strategies are lower-income nations where economic diversification remains a pressing task. For these countries, green hydrogen offers the possibility of getting in early on an emerging industry and using it to grow incomes and build stronger economies. A summary of hydrogen strategies from a selection of such countries, located in Africa and Latin America, is provided in Table 11.2 below.

Before we explore what's happening in these regions however, it is worth noting that wealth inequality between nations poses a risk as well as an opportunity. In addition to the common challenge of securing long-term buyers for green hydrogen, developing nations often face additional hurdles, including limited infrastructure, difficulty attracting finance, and higher borrowing costs when funding is secured. Consequently, in order to develop large scale industrial projects, they must rely on support from richer nations and multinational corporations, whose interests are not completely aligned.

The past offers no shortage of well publicised cautionary tales. In Nigeria's Niger Delta, decades of oil extraction enriched foreign corporations and political elites while leaving local communities with poisoned lands, shattered ecosystems, and chronic poverty. Likewise, in the Democratic Republic of Congo, cobalt and

Table 11.2. Hydrogen strategy summary: lower income export-focused nations.

	Strategy Document	Key Policy Instruments	Notable Targets
Egypt	November 2023; Green Hydrogen Strategy	- 10,000 km^2 land made available - Income tax credits (33–55%)	2030: 100 ktpa domestic 1.4 Mtpa export 2040: 2 Mtpa domestic 3.75 Mtpa export
Morocco	August 2021; National Hydrogen Development Strategy	- 10,000 km^2 land made available - Import and value added tax exemptions	2030: 120 ktpa domestic 300 ktpa export
Algeria	March 2023; Roadmap for the Development of Hydrogen	SouthH2 Corridor Collaboration	Supply around 10% of Europe's hydrogen by 2040
Tunisia	October 2023; National Strategy for the Development of Green Hydrogen and its Derivatives in Tunisia	Preliminary agreements with developers	8 Mtpa production by 2050 (6 Mtpa for export to Europe)
Mauritania	October 2023; Roadmap for the Development of a Low-carbon Hydrogen Industry in Mauritania	- 35 year operational licence - Suite of tax exemptions	Produce: 12.5 Mtpa by 2035 1% of global green H$_2$ and 1.5% green iron by 2050
Namibia	November 2022; Green Hydrogen Strategy	40 year land lease	Produce: 1–2 Mtpa by 2030 5–7 Mtpa by 2040 10–15 Mtpa by 2050
Chile	November 2020; National Green Hydrogen Strategy	$1 billion hydrogen development fund	Produce: 3 Mtpa by 2030 13 Mtpa by 2040 20 Mtpa by 2050
Argentina	September 2023; National Strategy for the Development of the Hydrogen Economy	Uncertain; law in limbo following change in government	Export: 300 ktpa by 2030 4 Mtpa by 2050
Brazil	August 2022; National Hydrogen Program	Production tax credits (value to be determined year by year)	No targets have been set
Uruguay	June 2022; Green Hydrogen Roadmap in Uruguay	Renewable energy tax incentives	2040: 500 ktpa domestic 1 Mtpa export
South Africa	October 2023; Green Hydrogen Commercialisation Strategy	- Strategic designation of hydrogen development zones - Green Hydrogen Atlas	Produce: 500 ktpa by 2030

copper mining provide raw materials for the energy transition while the miners themselves remain trapped in brutal conditions and surrounded by environmental devastation. There is concern that similar patterns could be repeated in the pursuit of hydrogen. Wealthy countries and corporations are once again seeking access to the land, sun, and wind of lower-income nations, aiming to ship clean energy home while the tangible benefits to local communities are of questionable merit.

In some cases, harnessing renewable power specifically for hydrogen may even be a misallocation of resources from the perspective of climate protection. Many of the countries that will be discussed – particularly in Africa – still rely heavily on fossil fuels for their electricity. More emissions would be prevented by simply replacing this generation with renewables, rather than making and moving hydrogen, while losing in excess of 30% of the energy in the process. Moreover, the scale of renewable energy required for hydrogen production can be enough to significantly decarbonize the supplier country's own grid. For instance, the engineering consultancy Ricardo has reported that Chile and Namibia would be able to completely decarbonize their electricity supply with the renewables required for their hydrogen plans.[4] This raises the uncomfortable prospect of clean energy being siphoned off to meet the climate targets of rich countries, leaving more effective and equitable options on the table.

These risks were recognised in a recent joint report from the UN, IRENA, and the German Institute of Development and Sustainability (IDOS), which argues that hydrogen development must be orientated around local needs. Developing nations should start small and build up, prioritizing decarbonization at home, rather than diving straight into mega projects that send clean molecules abroad.[5]

Nonetheless, the lure of export revenue is strong and the need for partnership with wealthy countries is real. Against this backdrop of opportunity and risk, lets now explore how countries across Africa and Latin America are positioning themselves to take advantage of the hydrogen economy.

Africa

According to the Energy Industries Council, 2025 began with 41 African green hydrogen projects in the works. These are spread across nine countries: Algeria, Djibouti, Egypt, Mauritania, Morocco, Namibia, South Africa, Tunisia and Zimbabwe.[6] All are in the early stages of development, with no final investment decisions having been taken.

Africa's growing interest in clean hydrogen has largely been the result of courting by Europe. In addition to European Commission representatives, leaders from Germany, France, Spain, Italy, Denmark, and the Netherlands have come calling in recent years, hoping to kick-start hydrogen exports in various African nations. In addition, more than $20 billion has been pledged towards the Africa-EU Green Energy Initiative, which is part of the wider EU Global Gateway programme. It is hoped that this will help fund at least 50 GW of renewable electricity, which is roughly the average demand of France, and support 40 GW of hydrogen electrolysis capacity, which would produce around 4 million tonnes per year.

North Africa is of particular interest because its close proximity to Southern Europe enables hydrogen transport via pipeline – a far more cost-effective option than shipping. In 2025, officials from Germany, Italy, Austria, Algeria, and Tunisia gathered in Rome to sign a declaration that they will jointly develop a hydrogen pipeline network to connect clean hydrogen projects in North Africa with demand centres in Europe. This Southern Hydrogen Corridor (SoutH2) is expected to be 4,000 km long, although 60–70% of it will use existing infrastructure. The hope is that it will eventually transport nearly 5 million tonnes per year, with Germany receiving a third of the supply. An additional route has been proposed between Morocco and Spain, which already share a natural gas pipeline. Morocco is also planning a 5,600 km long offshore natural gas pipeline that will run south to Nigeria, passing through the territory of 11 other countries. It has been suggested that a hydrogen pipeline could run in parallel, which would theoretically allow all these nations an export route to Europe.

Egypt is perhaps the most proactive African nation when it comes to clean hydrogen. While hosting the UN's COP27 climate conference in 2022, Egypt signed eight agreements to provide land for hydrogen developments within the Suez Canal Economic Zone, which includes numerous ports and industrial hubs. Many of these projects plan to take advantage of the strategic location by the Suez Canal to provide green shipping fuels. The proposed projects amount to 1.5 million tonnes of hydrogen production per year, broadly matching Egypt's 2030 target of 1.4 Mtpa. To support investment, the government is offering income tax credits ranging from 33% to 55%.

Egypt can also lay claim to having exported the continent's first green hydrogen. In 2022, the 15 MW Egyptian pilot facility shipped certified green ammonia to India, destined for the production of washing powder.

Moving further west, Algeria, which already exports natural gas via pipeline to Spain and Italy, aims to leverage its existing infrastructure to supply up to 10% of Europe's clean hydrogen needs by 2040. The government is conducting feasibility studies for a hydrogen pipeline under the Mediterranean, while its 2023 Hydrogen Roadmap envisions producing between 900,000 and 1.2 million tonnes of hydrogen annually. With the world's 10th largest natural gas reserves and good solar resources, blue and green hydrogen are both options.

Neighbouring Tunisia published its national hydrogen strategy in 2024, with help from Germany's international development agency (GIZ). Its stated goal is to produce 8 million tonnes per year by 2050, of which 6 million are intended for export to Europe via pipeline. The government also envisions using green hydrogen to decarbonize its fertilizer industry, which currently depends on imported ammonia.

Meanwhile, Morocco is targeting 120,000 tonnes of domestic consumption and 300,000 tonnes for export by 2030. As with Egypt, the government has allocated 10,000 km² of land for renewable hydrogen projects, as part of its 'Hydrogen Offer', which also includes tax incentives. Six projects have so far been awarded parcels of land.

If there is a Moroccan flagship project, it would be state-owned fertiliser producer OCP's planned one million tonne per year green ammonia facility. The company intends to use the ammonia to replace the fossil-based molecule in its operations. While acknowledging that green ammonia will cost more than grey, it sees this move as providing security of supply given that Morocco has no natural gas of its own and must import it from elsewhere. A 100,000 tonne pilot plant is planned as a first stage.

Morocco's southern neighbour Mauritania has also attracted large-scale green hydrogen proposals. In addition to plentiful land on which to build, and some of Africa's best solar and wind resources, Mauritania's 2024 Hydrogen Law offers developers a suite of tax exemptions along with 35-year operational licenses. Mauritania also has healthy reserves of iron ore, which has led the President of the European Commission, Ursula von der Leyen, to suggest that it also produce green iron for export to Europe. Major gigawatt-scale green hydrogen projects have been announced from companies based in Denmark, France, Serbia, and the UAE.

The most ambitious, and attention-grabbing, had been the Danish 35 GW Megaton Moon proposal. Dramatically set in the sun-scorched Sahara Desert, the developer's vision was to build 60 GW of solar panels and windmills in the shape of a crescent moon, in reference to that which is on the Mauritanian flag. This sprawling complex of glass and steel would span 60 km in diameter, an engineering marvel so vast it would be visible from the Moon. Unsurprisingly, this would have been among the largest green hydrogen complexes in the world, producing four million tonnes annually. However, in order to secure land rights, the proposal was subsequently scaled down by more than 80% at the request of the Mauritanian government, which appears to favour a more modest approach. This is an example of why any grand announcement of clean hydrogen production should be taken well salted.

Moving now to Africa's southern regions, Namibia's vast solar and wind resources have also attracted the attention of European investors. The EU has pledged €1 billion in public and private investment for Namibia's energy and infrastructure, including €25 million directly for hydrogen development. Additionally, Belgium's Port of Antwerp-Bruges is partnering in a hydrogen and ammonia export terminal located on Namibia's Walvis Bay. For its part, the Namibian government's plan is to make and export ammonia, e-methanol, synthetic kerosene and green iron. Namibia's Oshivela pilot plant, which began construction in 2023 is likely to produce Africa's first green hydrogen reduced iron, which it has agreed to sell to a German metals company.

Namibia's hydrogen ambition is large: it aims to pump out 10–15 million tonnes by 2050. The flagship project at the centre of this plan is the Hyphen Hydrogen Energy project; a 3 GW green ammonia facility planned inside Tsau Khaeb National Park. However, the proposal has drawn criticism from local environmentalists, who argue that placing the facility inside a national park threatens biodiversity and undermines conservation efforts. Some have questioned whether Germany— the presumed buyer of the hydrogen—would permit such a development within its own borders. This backlash highlights the ethical concerns associated with energy colonialism that were outlined earlier.

Like its neighbour, South Africa, has attracted strong interest from Europe as a potential green hydrogen supplier. Team Europe has offered €4.4 billion in finance, along with €300 million in grants to support the build out of renewable energy infrastructure, including green hydrogen. South Africa's Green Hydrogen Commercialisation Strategy notes that the country's books do not allow it to provide direct financial assistance so it will instead look to secure international partnerships such as that just mentioned. However, South Africa has been working to aid developers by identifying locations suitable for green hydrogen. A recently launched Green Hydrogen Atlas, maps the country on the basis of renewable energy potential and existing infrastructure. The Boegoebaai area, near the Namibian border, is of particular interest. Former state-owned energy and chemicals company Sasol has plans to develop a new deepwater port to support green hydrogen exports, alongside up to 40 GW of electrolyser capacity, and is seeking international partners to help move the project from concept to reality.

Meanwhile, across the Atlantic Ocean, another region rich in renewable energy is also drawing the attention of developers.

Latin America

Latin America already generates approximately 60% of its electricity from renewable sources – twice the global average.[7] This is largely driven by an extensive hydropower network, which has long provided a stable energy backbone. Now, its vast untapped solar and wind resources and large swaths of undeveloped land, make it an obvious target for green hydrogen development. This potential has drawn significant international interest, particularly from the European Union and development banks. From the powerful Patagonian winds to the extreme solar radiation of the Atacama Desert, Chile and Argentina offer some of the world's most favourable conditions for renewables. Similarly, the abundant hydropower and expanding offshore wind potential of Brazil and Uruguay are also very attractive. However, each nation is approaching the possibility of a hydrogen economy in a slightly different way.

Chile has been the most aggressive in its pursuit of a green hydrogen industry, launching a national hydrogen strategy in 2020 and updating it in 2024. Indeed, this strategy frames the export of clean energy as a new productive identity for a country that has historically had to import most of its energy. The strategy hopes to turn this on its head, stating that "green hydrogen and its derivatives represent a historic opportunity to transform Chile into one of the leading exporters of clean energy globally."

It's worth noting that fossil fuels still supply over a third of Chile's electricity, and the current goal is to reduce this to 20% by 2030. As noted previously, Chile would better serve global emissions reduction efforts by substituting new renewable electricity for existing fossil sources in preference to inefficiently exporting it as hydrogen-based fuels. Nevertheless, Chile's government is keen to push on with the later.

Citing analysis by the international consultancy McKinsey, the government's strategy document claims that Chile boasts 1800 GW of renewable energy potential, which for context is a little over half of China's entire electricity system. This would be enough to make a staggering 160 million tonnes of green hydrogen per year. Chile's actual targets are not so outlandish. By 2030, it hopes to nearly double its renewable electricity capacity and produce up to three million tonnes of hydrogen. Chile also projects, again based on McKinsey's analysis, that it could produce the cheapest hydrogen in the world, at less than $1.50/kg. However, this figure excludes transport costs, which would be significant given Chile's distance from key markets in Europe and Asia, meaning it may not be their lowest cost option overall.

To help achieve its hydrogen dream, Chile has established a billion dollar hydrogen development fund. The fund is largely made up of loans from various international investment banks: $400 million from the Inter-American Development Bank; $150 million from the World Bank; and around $110 million each from the European Investment Bank and the German Development Bank. The Chilean government topped up the pot with another $250 million, bringing the total up to 10 figures. It is hoped that this fund will make projects more bankable, and attract another $12.5 billion in private sector investment. By lowering the upfront risk to investors, the fund aims to accelerate the development of Chile's green hydrogen industry. As of 2025, this industry consists largely of the Haru Oni pilot plant, which was described in Chapter 4, and is producing 130,000 litres of e-fuel per year.

Despite being singled out by the IEA as having one of the world's cheapest locations for green hydrogen production, Chile's eastern neighbour has thus far shown less exuberance. The windswept plains of Argentine Patagonia are expected to produce green hydrogen at around $1.50/kg by 2030. Unsurprisingly, the EU has come calling, offering Argentina €200 million in funding for feasibility studies and project development.

Argentina's 2023 National Hydrogen Strategy had set an export target of 300,000 tonnes by 2030. However, just a few months after its release, the government was replaced with that of the fiery populist Javier Milei. The new administration had said that they would update the strategy in 2024, but as we entered 2025 this had still not happened. With Milei calling Argentina's natural gas reserves "a panacea" for its economy, and gas exports driving a record $5.7 billion energy surplus in 2024, it is unclear how much interest the government now has in hydrogen, and no major projects are currently in the works.

Energy superpower-in-waiting Brazil launched its National Hydrogen Program in 2022, aiming to decarbonize Brazil's economy, encourage technological development, and create a competitive market for hydrogen. This was somewhat curious given that Brazil's president at the time was Jair Bolsonaro – a vocal climate sceptic. However, the Ministry of Mines and Energy hinted at the lure of export dollars as motivation, noting that Brazil has great potential to stand out in clean hydrogen exporting since over 80% of its electricity comes from renewable sources. It's well known that the scale of Brazil's resources is vast, but the ministry has reportedly estimated its potential output of low carbon hydrogen at an astounding 1.8 billion tonnes, which is many times greater than predictions of global hydrogen demand by mid-century. Thus, Brazil has the potential to become a major producer.

In 2024, under the presidency of Lula da Silva, the National Low Carbon Hydrogen Policy was passed. In addition to creating a national certification scheme, this law also provides economic incentives for low carbon hydrogen, which Brazil defines as having less than 7 kg CO_2 per kg of hydrogen. This value is quite a lot higher than other countries because Brazil is a major producer of bioethanol from sugarcane, and the government clearly wants to ensure that hydrogen produced from ethanol has access to tax credits. However, the size of the tax credit awarded to a particular project will depend on the carbon footprint of the hydrogen, with more money awarded to cleaner projects. It is also expected that the 7 kg CO_2/kg H_2 will be reduced in the future.

The budget allocated for these hydrogen tax credits is \$3.4 billion, which will be distributed via annual competitive auctions over five years. Additional financial incentives include \$1.09 billion in funding (supported by the International Climate Investment Fund) for clean hydrogen projects that help decarbonise hard to abate sectors, and "REHIDRO", tax exemptions for the purchase of materials and equipment through to 2030. Brazil's hydrogen policy has so far not been terribly specific on how it intends green hydrogen to be used.

Southern neighbour Uruguay, with its 91% renewable electricity mix, is already one of the cleanest power producers in Latin America, placing it in a strong position for green hydrogen production. Its 2023 National Hydrogen Strategy highlights a substantial renewable energy potential, estimating that 90 GW of wind and solar resources could support the production of up to 6 million tonnes of green hydrogen per year. Uruguay also has significant biogenic CO_2 resources and iron ore reserves, which opens up the option of supplying green iron, methanol and e-fuels. Uruguay's green hydrogen roadmap envisions around half a million tonnes of domestic demand by 2040, mainly for heavy-duty transport, shipping fuels, and fertilizers, while an additional one million tonnes would be funnelled into exports, particularly in the form of ammonia, e-fuels, and aviation fuel.

Big Numbers: Chapter 11

- **7 million tonnes** - Total stated 2030 production targets for the high-income 'sellers'.

- **6 million tonnes** - Total stated 2030 production targets for the lower-income 'sellers'.

- **15 million tonnes** - Combined 2030 import targets for the 'buyer' nations. More than the above.

CHAPTER 12

Do It Yourself

Some countries have the resources to produce large volumes of clean hydrogen—and they intend to keep it. Although open to export, they are primarily working on using hydrogen to decarbonise their own industries. We call them the DIY (do it yourself) nations. A summary of their policies is provided in Table 12.1.

Table 12.1. Hydrogen Strategy Summary: DIY Nations.

	Earliest Strategy	**Notable Targets**	**Key Incentives**
China	March 2023; Medium and Long Term Plan for Hydrogen Industry	'Significant' use of clean hydrogen by 2035	National and Regional Targets
India	January 2023; National Green Hydrogen Mission	5 Mtpa by 2030	Subsidy for first three years: Year 1 - $0.60/kgH$_2$ Year 2 - $0.50/kgH$_2$ Year 3 - $0.40/kgH$_2$
United Kingdom	August 2021; UK Hydrogen Strategy	10 GW by 2030, with at least 50% green. (10 GW is about 5 Mtpa)	15 year CfD guarantees producers a price of £9.49/kgH$_2$ ($12/kgH$_2$)
United States	June 2023; US National Clean Hydrogen Strategy and Roadmap	Produce: 10 Mtpa by 2030 20 Mtpa by 2040 50 Mtpa by 2050	'45V' 10 year tax credits: $3/kg, < 0.45 kgCO$_2$/kgH$_2$ $1/kg, 0.45–1.5 kgCO$_2$/kgH$_2$ $0.75/kg, 1.5–2.5 kgCO$_2$/kgH$_2$ $0.60/kg, 2.5–4.0 kgCO$_2$/kgH$_2$

The United States

Emerging from the economic tarpit that was the covid pandemic, the passing of the Infrastructure Investment and Jobs Act (IIJA) was a signature achievement for President Joe Biden. Labelled as a "once in a generation investment", this 2021 legislation provided 550 billion dollars in new funding to modernise America's ageing infrastructure. This included an extra $62 billion for the Department of Energy (DOE) to distribute among clean energy projects, with $9.5 billion earmarked for hydrogen. This is clearly a healthy sum for hydrogen, although it's worth noting that the DOE's budget also included 2.5 billion for electric school buses, 7.5 billion for

DOI: 10.1201/9781003361428-15

building an electric vehicle charging network, and 73 billion to upgrade the power grid and facilitate renewable electricity expansion. Hydrogen was seen as part of a broader portfolio of electrification.

Further and even more significant support came in 2022 with the passage of the Inflation Reduction Act (IRA), which introduced the now-famous '45V' hydrogen production tax credit. This law provides a subsidy of up to $3 per kilogram of clean hydrogen produced, with the exact amount scaled according to emissions intensity. No budget ceiling was set, meaning the eventual cost could be enormous.

At the time it was announced, the 45V credit caused quite a stir internationally as it was the first scheme to subsidise hydrogen on a per-kilogram basis. On top of that, the amount of subsidy was – and still is – considered very generous indeed. Other nations with hydrogen aspirations worried that the US would suck up all the hydrogen investment dollars and make it difficult for them to meet their own production goals. A high-level task force was set up between the US and the European Union to iron out wrinkles in the IRA, in an attempt to avoid trade conflict. In addition to the EU, countries including India, Australia, and Canada began introducing their own hydrogen production incentives out of concern that they would be left behind. It's fair to say that the IRA remains one of the most influential additions to global hydrogen policy.

Interestingly, the first US hydrogen strategy wasn't published until 2023 – a good while after the money to support it had already been announced. It listed climate, energy security, energy resilience, and the opportunity to create economic value as motivating factors, and prescribes three broad courses of action.

The first is to focus on strategic, high-impact uses for hydrogen such as industrial processes, heavy transport, and long-duration energy storage. Exporting hydrogen and its derivatives was seen as a secondary goal once domestic needs had been met.

The second goal is to drive down the cost of producing clean hydrogen. In a play on the term 'moonshot'—President Kennedy's famous pledge in 1961 to land a man on the moon within a decade—the 'hydrogen shot' was announced in mid-2021. Its aim is to bring the price of hydrogen down to $1 per kilogram within ten years, the so-called '111' target. This is the price where unsubsidised hydrogen will become economic for most industries.

The third pillar of the strategy was to support the development of hydrogen hubs. In 2023, seven regional hubs were awarded a share of $7 billion from the IIJA hydrogen pot to establish large-scale production, distribution, and end-use systems across the country.

The US government's support for clean hydrogen generated a surge of optimism in the industry, but this quickly gave way to frustration as the tax credits floundered amid regulatory delays. The US Treasury had been given the job of deciding the rules around tax credit eligibility, and it took nearly one and a half years to make the first draft available for consultation. Nearly 30,000 responses were received, reflecting the amount of money that was up for grabs, as well as the law's potential impact on greenhouse emissions. The 'three pillar' rules, modelled on strict European standards, proved particularly contentious. This level of interest contributed to yet another year of delay, with the final rules—only a slightly softened version of the draft—published in January 2025. This was in the final days of Biden's presidency and just before the start of Donald Trump's second term.

Trump has frequently expressed scepticism about climate change, going so far as to call it "a hoax." During his 2024 presidential campaign, he railed against his predecessor's environmental initiatives, referring to them as the "Green New Scam" and pledging to rescind any unspent funds from the IRA and IIJA. It was therefore not surprising when a flurry of executive orders on his first day back in office did just that. Other orders included pulling the US out of the UN climate accord (again), halting government approvals for wind energy projects, and eliminating federal incentives for electric vehicles. The direction of travel under the new administration is clear: the clean energy transition will not find an ally in Donald J. Trump.

For the hydrogen industry in particular, the freeze on IIJA disbursements was of real concern because it included the $7 billion allocated to hydrogen hubs. Reporting had suggested that the hubs were at risk of having their funding revoked, although optimism remained that blue hydrogen projects – located in states traditionally supportive of Trump – would maintain their support. Known for his "drill, baby, drill" catchphrase, Trump is a staunch backer of oil and gas interests, so it is anticipated that he will be amenable to rules that aid blue hydrogen producers, regardless of whether greenhouse emissions go up or down as a result. Months later, the Secretary of Energy clarified that hub funding is not at risk, but the unease caused by the interim uncertainty was real.

It was initially believed that the 45V tax credit was on firmer ground as it is an entitlement enshrined in tax law – not a pool of money that can be frozen by executive order. Further, 45V has received strong support from a formidable lineup of American businesses, environmental groups, and politicians from both major parties. This unusual coalition is working to reframe clean hydrogen in terms that align with Trump's "energy dominance agenda": highlighting its role in energy security, industrial competitiveness, and export potential. The need to keep pace with China's growing dominance in hydrogen technology has also been highlighted as of particular concern. However, by mid 2025, legislation to scrap 45V was reportedly on the table, adding further uncertainty for the clean hydrogen industry.

It must be noted that maintaining production incentives is only one side of the coin. Although incentives for users of hydrogen were discussed under Biden, they were never implemented, and are very unlikely to feature under Trump. This matters because, as reported by *S&P Global Commodity Insights*, industrial users in the US have shown little interest in paying more than the $1/kg they currently pay for fossil hydrogen. Without a strong domestic market, Trump's America may be more seller than DIY—relying on exports to Europe and Asia to make large-scale hydrogen production viable. But here too the political winds are unfavourable. A turn toward protectionist policies and tariff wars could easily make American hydrogen uncompetitive abroad. Given how marginal the economics of clean hydrogen already are, even modest trade barriers could derail the export business case.

In its 2025 hydrogen update, the DOE reported that it now expects significant project delays and cancellations due to political uncertainties and increasing costs. It forecasts that 2030 clean hydrogen production will now be in the 7 to 9 Mtpa range—well below earlier expectations of 14 Mtpa, and missing the national target of 10 Mtpa.[1] America's clean hydrogen path, once paved with optimism, now looks increasingly winding and unstable.

China

China is by far the world's largest producer and consumer of hydrogen. In 2023 it used 28 million tonnes, which was more than double that of the next largest user – the US.[2] Its capacity to produce hydrogen is also growing rapidly; nearly doubling from 22 million tonnes in 2018 to 40 million tonnes in 2021. This hydrogen is mostly used as a feedstock in chemical industries and over 75% of it is currently produced from the gasification of coal.[3]

However, the share of hydrogen produced from fossil fuels is expected to decline, as China rapidly expands both renewable power and electrolyser manufacturing. In 2020, it held less than 10% of global electrolyser production capacity; by 2025, that figure had risen to over 60%. This surge is significant not only for domestic use but also as an export stream—with China expected to supply more than a third of electrolyser orders outside of the US and Europe, according to energy consultancy Wood Mackenzie.

Hydrogen energy, and fuel cell vehicles in particular, have been part of China's research agenda since the 2000s. While the Chinese government does not release data on public investment, it is estimated from research institution statistics that as of 2019, China had spent around 20 billion yuan ($2.9 billion) on developing fuel cell vehicles and 10 billion yuan ($1.5 billion) on other hydrogen technologies.[4] It is reasonable to assume that this level of spending will continue if not increased. The most recent five-year plan for China's energy system (2021–2025) states that research and development funding for energy will increase at an average of 7% per year. This plan also lists hydrogen as one of the six "industries of the future", along with artificial intelligence (AI); quantum computing; semiconductors; biotechnology; and new energy vehicles (EVs).

The primary motivation for China's interest in hydrogen energy appears to be decarbonisation, along with developing new industries. China is somewhat unique among the countries discussed here in that its government is a permanent fixture, capable of setting and pursuing long-term policy objectives without the disruptions or reversals that can accompany electoral cycles. While there are no doubt substantial flaws with this form of government, its consistency may offer advantages in the case of energy transition and decarbonisation.

In September of 2020, the Chinese government announced that China's CO_2 emissions would peak by 2030 and that it would achieve net-zero emissions by 2060. Around the same time, the government introduced a scoring system with financial incentives to encourage cities and regions to build hydrogen value chains and tackle technical challenges. By 2021, 32 cities and 9 provinces had received the message and released five-year plans for hydrogen. The central government's *Medium and Long-term Plan for the Development of the Hydrogen Industry (2021–2035),* was published in early 2023. It was supposed to have be released in 2020, and this delay suggests a degree of caution and uncertainty within the Chinese government around the role of hydrogen energy in decarbonisation. Further, the plan does not have a lot of firm targets, and is vaguely descriptive in nature. The stated goals are as follows:

By 2025, the core technology and manufacturing processes for hydrogen and fuel cells will be basically mastered. There will be 50,000 fuel cell vehicles operating and an (unspecified) number of hydrogen fuelling stations built. Renewable hydrogen production capacity will be between 100,000 and 200,000 tonnes per year. By 2030, the clean hydrogen supply chain will be more or less complete and by 2035 hydrogen will be used across numerous applications and the proportion of clean hydrogen used will be significantly increased.

There are a few notes to make here. Firstly, adding up all the targets from the regional and city hydrogen plans results in more than 60,000 fuel cell vehicles and 550 hydrogen fuelling stations. The central government's plan is more conservative than this, which gives the impression that it is trying to temper the enthusiasm of regional governments and pump the metaphorical brakes on hydrogen vehicle roll-out. To provide some context, China already had around 7000 hydrogen vehicles on the road in 2020 (along with 4 million battery electric vehicles).[5] Interestingly, these hydrogen powered vehicles were either commercial trucks (60%) or buses (40%). Only a tenth of one percent of China's hydrogen powered vehicles in 2020 were passenger cars.[6]

Secondly, the hydrogen production target of 100,000 to 200,000 tonnes by 2025 is also very conservative, especially given that China's green hydrogen capacity may have already reached 500,000 tonnes in 2021.[7] This cautious stance reinforces the impression that China is taking a pragmatic, step-by-step approach: refining technical capabilities and trailing energy systems before scaling up. However, while the pace may be deliberate, the groundwork for hydrogen expansion is still being laid. In November 2024, China changed the official classification of hydrogen from a 'hazardous chemical' to an 'energy resource'. This will reduce red-tape and make it easier to apply hydrogen in non-traditional use-cases.

Around the same time, the influential National Reform and Development Commission released a policy document titled *Guiding Opinions on Vigorously Implementing Renewable Energy Substitution Actions*. It sets out a broad vision for the energy transition, and like the hydrogen plan, is short on specifics and offers few numerical targets. One exception is the goal to increase renewable energy consumption from the equivalent of 1.1 billion tonnes of coal in 2025 to more than 1.5 billion tonnes by 2030.[8] That final figure translates to around 30 EJ – roughly three times the annal electricity demand of the European Union.

What the *Guiding Opinions* makes clear is that the authorities want China to try a bit of everything, calling for the "active and orderly development" of solar PV, solar thermal, onshore and offshore wind, hydropower, biomass, geothermal, biofuels—and hydrogen. It also gives some indication of where hydrogen is most likely to be used. The document points to large-scale substitution for grey hydrogen in the production of ammonia, methanol, petrochemicals, and steel. It mentions the promotion of hydrogen metallurgy including the direct reduction of iron with hydrogen. In transport, support is encouraged for pilot projects in shipping and aviation that use green hydrogen, ammonia, and biofuels—including sustainable aviation fuels, which depend on hydrogen for their production.

Governors are also directed to "Explore the construction of integrated bases for wind, solar, hydrogen, ammonia, and alcohol", and "Scientifically guide the orderly transfer of industry to areas rich in renewable energy". Building hubs of complementary industries in renewable energy rich regions is a recurring theme throughout the document.

Despite transport being the early focus of Chinese hydrogen research, the guidance suggests that electrification of transport is now the priority. It states that China should "Accelerate the development of electric passenger cars, steadily promote the replacement of buses with EVs and explore the promotion and application of new energy medium and heavy trucks". Presumably hydrogen is a 'new energy medium' that should still be explored for heavy trucking. It is clear that hydrogen fuel still has a place in transport, as the advice is to strengthen the transport infrastructure, including gas stations, hydrogen stations, and both the urban and rural charging network.

The guidance document is also illuminating for what it doesn't say. Hydrogen is noticeable by its absence in the section on heat for buildings, which presumably means that China does not have any interest in this use case.

According to analysis from the Orange Research Institute in Shanghai, as reported by *Hydrogen Insight*, the green hydrogen developments in China's project pipeline align with the *Guiding Opinions*. The chemicals industry is expected to consume 90% of the green hydrogen that will be produced. Almost half of this will be for methanol production, with ammonia also taking a sizable slice at 31%. Smaller volumes will be directed to making jet fuel (9%), and oil refining (7%), with the remainder used for miscellaneous chemicals. The 10% that is not earmarked for chemicals will be distributed more or less evenly across transport fuel, energy storage and power generation, and other industries – particularly metals. Thus, the alignment between the Chinese government's hydrogen policy goals and what appears to be happening on the ground is, perhaps unsurprisingly, notably strong.

The United Kingdom

In contrast with its former colony on the other side of the Atlantic Ocean, both major political parties in the UK broadly support the development of clean hydrogen. The UK's first hydrogen strategy was published in 2021 under the Boris Johnson Conservative government, and has been updated annually since. The successor Labour government has called hydrogen "a crucial enabler" of a low carbon energy system, and one that can contribute to energy security, decarbonisation, and job creation.

Creating a low carbon energy system is high on the agenda for Labour. With around 40% of its energy coming from natural gas, the UK was hit particularly hard by the price spike resulting from Russia's invasion of Ukraine. The Government therefore views the energy transition as a way to simultaneously tackle economic wellbeing, climate obligations, and national security. The goal of Labour's flagship climate policy – the *Clean Energy Superpower Mission* – is to have 95% of the UK's electricity generated from low carbon sources by 2030.

To help enable this shift, the previously separate gas and electricity system operators have been merged into a single independent body: the National Energy System Organisation (NESO). With a holistic view across the entire energy system, NESO leaves the UK better placed to make rational choices around decarbonising power, heating, and transport. In the case of hydrogen, this means deploying it only where it's genuinely needed—not where entrenched interests might once have pushed it.

As to the question of where hydrogen is genuinely needed, the answer, which was once broad and speculative, is now becoming confident and focused. In early 2025, the UK's influential Climate Change Committee (CCC)—the government's independent climate adviser and the body responsible for setting legally binding emissions limits—published its *Seventh Carbon Budget*. This report describes hydrogen as playing a "small but important" role in the future energy system, best used to provide high-temperature industrial heat, act as a feedstock for synthetic aviation and shipping fuels, and serve as a form of long-duration energy storage to complement renewable power. This view is broadly shared by NESO, which emphasises hydrogen's value in sectors where electrification is not practical — rather than as a blanket solution.

The CCC is also clear about where hydrogen should not be used. On the UK's ongoing dalliance with hydrogen for home heating—which successive governments have acknowledged as "an area of extensive debate" without being willing to commit either way—the CCC has urged a definitive rejection, warning that delay only undermines investment in efficient electric alternatives. It also recommends abandoning plans to blend hydrogen into the gas grid, calling this a costly and ineffective way to create early hydrogen demand. In transport, the CCC foresees no role for hydrogen in cars or vans, and at most a limited role in heavy goods vehicles.

The CCC now expects hydrogen demand to reach around 60 TWh by 2040, which is approximately 1.8 million tonnes.[9] That's a sharp departure from its 2020 projections, which assumed nearly three times as much hydrogen would be needed to decarbonise buildings, transport, and industry. With those expectations now trimmed back to focus on harder-to-electrify sectors, the revised estimate lands surprisingly close to the Government's target of 10 GW production by 2030. Assuming electrolysers are running a good amount of the time, 10 GW should get you about 1 Mtpa. Rather than revealing a gap between ambition and planning, as seen in many other countries, these numbers suggests the UK is converging around a more grounded vision of hydrogen's role.

To help support its hydrogen ambitions, the UK has introduced a mix of subsidies, mandates, and direct public investment. Top of the list is £25 billion for subsidized production. Under a contracts-for-difference-style scheme, selected producers are guaranteed a price of £9.49 ($12) per kilogram. The early round awarded support to 11 green hydrogen projects with a combined capacity of 125 MW – small facilities by global standards.

Mirroring the EU, the UK has also introduced a Sustainable Aviation Fuel (SAF) mandate, which requires that 2% of jet fuel on departing flights comes from sustainable sources. This fraction will increase over time, requiring an increasing supply of hydrogen, which many SAFs require as a feedstock.

Somewhat controversially, the UK is also investing heavily in blue hydrogen. Two large carbon capture and storage (CCS) projects in northern England are set to receive £22 billion in public funding, and are expected to include a combined 100 kt per year of blue hydrogen production.[10] This approach has drawn criticism from the non-profit think tank Carbon Tracker, which warns that it could lead to higher overall emissions. Beyond broader scepticism about the climate integrity of blue hydrogen, their analysis points out that the natural gas required for production would likely need to be replaced in the near term by imported LNG—which carries a significantly larger upstream carbon footprint. With the UK currently producing only 40% of its own natural gas, it's argued that pursuing blue hydrogen is at odds with the stated goals of energy security and self-sufficiency.

Even so, the UK's hydrogen policy stands out for its relative coherence and pragmatism.

India

India is the world's most populous country and its largest democracy. It also has the fastest growing major economy and is expected to overtake Germany and Japan for third spot on the leaderboard in the coming years. This will involve a massive increase in energy use—perhaps by as much as a third over the next decade.[11] Currently, India imports over 40% of its primary energy requirements, costing it more than $90 billion annually.[12] Furthermore, 75% of India's primary energy consumption in 2022 came from fossil fuels, contributing nearly 7.5% of global energy-related emissions.[13] At the same time, the cost of solar and wind power in India is among the lowest in the world.

Within this context, India has announced the twin goals of energy independence—known as *Aatmanirbhar*—by 2047, and net zero by 2070. The net zero target is 20 years later than the global consensus of 2050, but India argues that it is still developing, with hundreds of millions yet to be lifted from poverty. Substituting locally produced clean energy for imported fossil fuels clearly offers both economic and environmental benefits, and in early 2023 India added hydrogen to its energy policy with the adoption of its Green Hydrogen Mission.

The stated objective of the Mission is highly aspirational:

> "To make India the Global Hub for production, usage and export of Green Hydrogen and its derivatives. This will contribute to India's aim to become Aatmanirbhar through clean energy and serve as an inspiration for the global Clean Energy Transition. The Mission will lead to significant decarbonisation of the economy, reduced dependence on fossil fuel imports, and enable India to assume technology and market leadership in Green Hydrogen."

India is the world's second-largest consumer of fertiliser, which requires large volumes of ammonia, for which India is already the third-largest producer. Ammonia is therefore a natural use case for green hydrogen. The same applies to steel, where India again ranks second globally, behind only China. Other applications highlighted by the Mission include refining, heavy-duty transport, marine fuels, and—more controversially—blending with natural gas.

In addition to becoming a leader in the production and use of green hydrogen, India aims to be a major player in manufacturing electrolysers and other enabling technologies. It has been widely reported that energy analyst Rystad Energy expects India's electrolyser manufacturing capacity to reach 8 GW per year by 2025: around 20% of the IEA's projected global capacity.

India's current goal is to produce 5 million tonnes of green hydrogen annually by 2030, which is roughly equivalent to its current hydrogen consumption. The strategy also allows for the possibility of increasing this to 10 million tonnes if export markets develop sufficiently.

The first step toward this goal was taken in January 2025, when construction began on the country's inaugural hydrogen hub in Pudimadaka, on the eastern coast. This green energy complex is expected to include 20 GW of renewables and produce nearly 550,000 tonnes of green hydrogen annually, to be used in fertiliser, methanol, and sustainable aviation fuel production.

The Green Hydrogen Mission was released with an initial outlay of approximately $2.4 billion in financial support. Around two-thirds of this is allocated to subsidising green hydrogen production. Subsidies are limited to the first three years, with maximum payments of around $0.60, $0.50, and $0.40 per tonne in years 1, 2, and 3. This is modest compared to the maximum amounts on offer elsewhere, but aligns with what was actually awarded in the EU Hydrogen Bank's first round.

Smaller pots of money—ranging from $50 million to $500 million—have been allocated to support pilot projects in electrolyser manufacturing, shipping, green steel, and refuelling infrastructure. The Mission also allows for consumption mandates, though the government has flip-flopped on whether to implement them.

As always, the devil is in the details, and India's green hydrogen policy is not without controversy. Green hydrogen is defined as having a well-to-gate emission of no more than 2 kg CO_2/kg H_2, including water treatment, electrolysis, gas purification, drying, and compression. While this is relatively stringent, the effectiveness of the policy's three-pillar guardrails is questionable. Producers are allowed to time-match their hydrogen production with renewable electricity on an annual basis—despite 70% of India's grid being coal-powered—which could result in electrolytic hydrogen with a carbon footprint higher than that of grey hydrogen from methane.

Policy direction is further muddied by India's Ministry of Coal, which is providing over $700 million in funding for coal gasification as a hydrogen production method. With emissions of 25 kg CO_2/kg H_2, coal gasification is the most polluting hydrogen pathway. Making matters worse, the quality of India's domestic coal is poorly suited for gasification, meaning imported coal would likely be needed—undermining both the climate and energy security goals.[14]

Despite these contradictions, investment interest remains strong. It has been widely reported that energy analyst BloombergNEF predicts only India and China will be able to produce green hydrogen at costs comparable to grey hydrogen without subsidies by mid-century. With enthusiastic government backing, vast domestic demand for clean energy, ultra-low renewable electricity prices, large-scale electrolyser manufacturing potential, and a strategic push for energy independence, India is now widely viewed as one of the world's most promising green hydrogen markets.

Big Numbers: Chapter 12

- **$3** - Per kilogram green hydrogen subsidy offered by the US under Biden – so generous it spooked the world. Delays in finalising the rules and possible Trump-era reversals may mean it's never paid out.

- **60%** - China's share of electrolyzer manufacturing capacity in 2025 – a further step in its march to clean tech dominance.

- **10 GW** - The UK's 2030 target for low-carbon hydrogen production, which is around 1 Mt per year – a figure well aligned with projected consumption of 1.8 Mt by 2040.

- **5 million tonnes** - The upper end of India's production target range for 2030 – roughly equivalent to its current hydrogen consumption.

Phantom or Panacea?

Long ago, in the time of the ancient gods, a war of succession raged between the Titans and their children: the Olympians, led by Zeus. Blessed with the power of foresight, our hero Prometheus foresaw the defeat of his tribe and together with his brother, abandoned the Titans and allied themselves with Zeus. For this, they were rewarded with the task of overseeing the creation of Earth's creatures.

His brother Epimetheus bestowed the beasts with gifts that would allow them to thrive: keen senses, powerful muscles, and sharp claws. Prometheus, meanwhile, set to work moulding man out of clay but by the time he was done, his brother's bag of goodies was empty – leaving humanity weak and defenceless. Unwilling to accept this fate for his creation, Prometheus stole for them the greatest force the gods possessed—fire.

The Romanian-American economist Nicholas Georgescu-Roegen later borrowed this Greek myth to describe energy technologies that are so powerful, they transform civilization. In his view, the ability to control fire was the first "Promethean Technology" because it enabled early humans to reshape their environment, extend their range, cook food, forge tools, and ultimately build the foundations of complex societies.[1] The second Promethean technology, according to Georgescu-Roegen, was the heat engine. Beginning with the steam engine and evolving into the internal combustion engine, gas turbine, and jet engine, this technology has powered the material growth of the modern world since the Industrial Revolution of the 18th century.

'Promethean' does not imply 'sustainable', however. These technologies still require an energy source—be it wood, coal, or oil—and are thus constrained by physical limits, such as the availability of fuel or the capacity of the biosphere to absorb their wastes. It is the latter which is the heat engine's Achilles' heel. The continuous release of carbon dioxide gas can no longer be tolerated due to the dangerous influence it has over Earth's natural systems.

The defining challenge of our time is to create a new technological base, capable of supporting industrial society without undermining the biosphere—to replace the heat engine with something *clean*.

Electrons First

At its heart, the transition to a sustainable energy system is a quest to elevate the electric motor. Versatile, emission-free, and ultra efficient. An elegant engine, for a more civilized age. Only through electrification can we replace polluting fossil stocks with flows of clean energy to fuel our societal metabolism.

DOI: 10.1201/9781003361428-16

Alas, being best available is not the same as being perfect. Electric devices require a physical connection to an electricity source via cables and wires, which can be awkward and expensive to build. For those devices that cannot remain stationary, batteries offer a solution, but these are also expensive and hold much less energy than the fossil fuels to which we are accustomed. At the system level, electricity must be used the moment it is generated. Balancing supply and demand is technically complex and requires extensive, often expensive, infrastructure. It's also impractical to transmit over long distances, limiting our ability to trade.

Enter hydrogen – the energy carrier.

Hydrogen can drive an electric motor via a fuel cell. It can also be burnt to fuel a heat engine – without releasing CO_2. Straddling past and future, it appears to offer the best of both worlds—the familiarity of combustion, the promise of electrification. This perceived versatility has seen hydrogen promoted as a silver bullet: a replacement for fossil fuel in just about every application.

However, unlike fossil fuels, which come in a ready-to-burn form straight out of the box, hydrogen is not an energy *source*. The hydrogen business is one of conversion, not extraction. At every conversion step, some energy is lost, with the end result that we get out less than we put in. Hydrogen production is, in the words of Georgescu-Roegen, a 'parasitic technology'. Such technologies can be clever, maybe even vital, but ultimately are dependent on a foundational Promethean tech that can provide excess energy. Hydrogen cannot support civilization on its own.

That is, unless geologic hydrogen proves a viable resource. However, we have not yet proven there are large scale reservoirs of retrievable hydrogen underground, let alone shown they can be extracted economically. Banking on that is like basing one's retirement plan on winning the lottery.

So, while we wait for our numbers to come up, we're left with hydrogen we know how to make. There are several options, but the climate benefit of hydrogen rests heavily on how it is made. Grey hydrogen is more polluting that the fuel it would replace, so that's out. Blue hydrogen has an attractive narrative, but natural gas leakage remains a threat, and underground storage of CO_2 has proven unpredictable, making it an unreliable option for widespread use. Turquoise hydrogen could be better, but its usage will be limited by the mountains of solid carbon that come with it.

The only truly clean hydrogen is 'green'; made through electrolysis powered by renewable electricity. Unfortunately, this process is costly, in both dollars and kilowatt hours. Electrolysis consumes about 30% of the input energy. Squeezing it into a form compact enough to store or transport costs another 10–30%. When it comes time to use the energy, nature's tax collector comes calling again. Burning hydrogen in an engine wastes around 70% of the energy, while fuel cells lose 40–50%. Hydrogen may be an energy carrier, but it's an awkward one.

As a result, technology that is powered directly by electricity is always a lot more efficient than those that employ hydrogen as an intermediary. Battery electric vehicles are 2.5–3 times more efficient than an equivalent hydrogen fuel cell vehicle. Heating a building by burning green hydrogen requires 4–6 times more renewable energy than an electric heat pump. Storing electricity in lithium-ion batteries typically retains over 80% of the original energy for later use. Using hydrogen for the same purpose provides around 30%. No matter how expensive electrification is, hydrogen will always be more so—simply because it takes more energy to achieve the same result. It pays to cut out the middleman.

Molecules that Matter

Hydrogen may not be suited to every task, but that doesn't mean it has no role. In fact, we've been working with it for a century, although it has historically been valued for what it enables as a molecule, not for the energy it holds. In this role, it is not a fuel but a reagent—a means of making something else. The benefit isn't just in the energy it carries, but in the chemical transformations it enables. Those applications will remain and are likely to grow.

Hydrogen is a vital ingredient in the production of ammonia fertilizer, which in turn has become essential for sustaining our swelling population. Very soon, ammonia may also be powering our large ocean-going ships, which require far more energy to move than batteries can supply. When made by combining air-captured nitrogen with electrolytic hydrogen—all powered by renewable electricity—ammonia becomes a near-zero-carbon fuel. Another leading shipping fuel candidate is e-methanol, and it too counts green hydrogen as a core ingredient.

Similarly, aircraft require an energy dense power source that batteries simply can't provide. The hydrogen molecule isn't much better in this regard, but it can be used to manufacture synthetic versions of the fuels we use now – offering a route to reduce, if not eliminate the life-cycle emissions from flying. While not a perfect solution, it may be the best option we have.

Arguably, the most exciting new hydrogen application, and the one with the greatest potential for emissions cuts, is steel production. Hydrogen can replace coal and natural gas as the chemical agent responsible for liberating iron from ore. Indispensable but highly polluting, steelmaking is responsible for nearly 8% of global emissions. Hydrogen offers a path to zero-carbon steel without compromising quality.

None of these applications come cheap—but hydrogen is justified not by efficiency, but by necessity. Where electrification isn't feasible, molecules must do the job, and as these examples show, some of our most vital industries still fall into that category. In these cases, the goal is not to beat electricity, but to replace fossil molecules – and for that, green hydrogen is often indispensable.

The Weight of Scale

Imagining a decarbonised world is one thing. Building it is another.
International Energy Agency's scenarios suggest that global electricity generation by mid-century will be between two and three times greater than it is now. Most of this growth will come in the form of renewables, which are now the cheapest option in many cases. The more we electrify with clean generation, the closer we get to net-zero, but delivering this growth is a monumental task.

The numbers are sobering.

The estimated cost for the renewables buildout is \$50–80 trillion, or around 2% of global GDP per year. The land requirement could be between half a million and 1.35 million square kilometres for wind and solar installations—an area comparable to Spain at the low end, and Peru at the high end.

A healthy slice of this capacity would be needed for hydrogen. Even if we reserve green hydrogen for essential uses – steel, fertilizer, and fuels for shipping and aviation – annual demand could reach 400 million tonnes by 2050. While this is probably an extreme value, it would require electricity equivalent to two thirds of global generation in 2023. It would also require building new handling infrastructure such as pipelines, port facilities and ships, not to mention 800,000 plus new electrolyzer stacks to make the hydrogen in the first place.

All this cost results in expensive green hydrogen, which can easily reach six times the price of conventional grey hydrogen, and even double that in resource-poor locations. With a gap this large, there is no way a market for green hydrogen will develop organically. Government support is essential, but it must also make sense.

Political Dreams and Reality

The latest round of hydrogen-enthusiasm has seen governments try just about everything to get the product flowing: tax credits, grants, price guarantees, infrastructure spending, and diplomatic outreach. The US is offering \$3 per kilogram of green hydrogen. The EU is mandating hydrogen-derived fuels and auctioning billions in subsidies. India, Australia, and Canada all have their own flavours of subsidy too.

And still, only a small percentage of projects have moved from conception to commitment.

Undoubtedly, boosters of the clean hydrogen industry are frustrated that it has not developed as rapidly as many had hoped, but the fact is, the cost of green hydrogen is still too high, and except for a few trailblazers, buyers are not prepared to pay the price. Without guaranteed buyers, expensive process plants don't get built. It's that simple.

As we wait for prices to fall, what hope there is of getting the industry moving in the near-term rests almost entirely on sustained government (and taxpayer) support. While this support has been plentiful, it's often influenced more by politics than by science. The EU, for example, currently uses around 7 Mt of hydrogen per year, and its fuel mandates are expected to add just 3.5 Mt more—making its 2030 consumption target of 20 Mt look wildly out of step. Meanwhile, Japan and South Korea have both made imports of hydrogen-based energy carriers a key part of their energy strategies – apparently locking themselves into high energy prices. And in the UK, hydrogen heating has become a political football—a distraction that's slowed the rollout of more efficient electric heat pumps. The political allure of hydrogen has often outweighed the simple logic of cutting out the middleman.

Furthermore, governments that design support packages will not necessarily be around to honour them. Projects that take a decade to build require confidence that the rules won't be rewritten halfway through. America's 2025 policy whiplash is a case in point—and with the possible exception of China, no country is immune.

Sustainability Under Scrutiny

Beyond political instability, there's also the question of whose needs are being served. Several wealthy countries—particularly in Europe, Japan, and South Korea—hope to meet their hydrogen targets by importing from nations with abundant solar and wind resources. Many of these prospective exporters are lower-income countries, often located in water-stressed regions. Critics argue that it would be better to focus new infrastructure on local development needs, rather than the energy demands of the Global North. While outside investment can help fund such projects, without strong safeguards, hydrogen exports risk reviving old extractive patterns—the newest reincarnation of colonialism.

There's also the small matter of climate integrity, which is after all the reason we're discussing hydrogen at all. It is not well appreciated that hydrogen molecule is an indirect greenhouse gas, as well as inherently prone to leaking. Ramped up production and widespread trade could see hydrogen become a double agent in the climate fight.

In addition to the climate impact of hydrogen itself, we have the carbon footprint of making it. Governments want to support clean hydrogen—but defining "clean" has proved contentious. The EU and U.S. tried to ensure integrity through the "three pillars" but these rules have met fierce resistance from industry, who argues they make projects too expensive to build. On the other hand, if the rules are too weak, we could end up subsidizing increased emissions.

In short, we're trying to build a new global energy system on a foundation of uncertain economics, contested definitions, and volatile politics. That's not easy.

This isn't to say that nothing is happening. The first major developments are starting to come online. Perhaps it isn't so much that progress has been slow—a building program this large was always going to take time. Perhaps the hype simply lifted expectations too high. The bold commitments of clean-hydrogen pioneers, flawed as they may be, have helped kick-start an industry that may eventually mature, but the scale of what's needed is breathtaking. So too are the challenges.

A Supporting Role, Not the Lead

Throughout this book, we've examined hydrogen's strengths and weaknesses—its chemistry, its costs, its promise, and its politics. What began as a bold vision has been narrowed by thermodynamics, economics and practicality. What remains is not the limelight, but a modest role in a much larger cast.

So, to the titular question: is hydrogen a sustainable energy panacea?

Certainly, it is not. Hydrogen's physical properties make storing and moving it too cumbersome. The energy lost in making and using it is too great. As an energy carrier, it is a phantom solution.

Hydrogen is not a gift from Prometheus. It will not do for the 21st century what fossil fuels did for the 20th. Used wisely it can help us get to net zero, but we must avoid using it where it can't.

We don't have the time, the money, nor the electrons to waste.

References

Chapter 1

1. Erisman, J.W., Sutton, M.A., Galloway, J.N., Klimont, Z. and Winiwarter, W. 2008. How a century of ammonia synthesis changed the world. Nat. Geosci. 1: 636–639.
2. Dawson, V.P. and Bowles, M.D. 2004. Taming liquid hydrogen: The Centaur upper stage rocket, 1958–2002 (NASA SP-2004-4230). NASA History Division.
3. Bockris, J.O'M. and Appleby, A.J. 1972. The hydrogen economy: an ultimate economy? Environ. This Month 1: 29.
4. Romm, J.J. 2004. The Hype About Hydrogen: Fact and Fiction in the Race to Save the Climate. Island Press.
5. U.S. Department of Energy. 2017. Historical DOE Hydrogen and Fuel Cell Funding (1975–2017) (DOE/EE-17006). Office of Energy Efficiency and Renewable Energy.
6. Swain, M.R. and Swain, M.N. 1992. A comparison of H_2, CH_4 and C_3H_8 fuel leakage in residential settings. Int. J. Hydrogen Energy 17(10): 807–815.
7. Esquivel-Elizondo, S., Shen, L., Dubovikoff, T. and Michanowicz, D.R. 2023. Wide range in estimates of hydrogen emissions from infrastructure. Front. Energy Res. 11: 1191098.
8. U.S. Department of Energy. 2009. Energy Requirements for Hydrogen Gas Compression and Liquefaction as Related to Vehicle Storage Needs (DOE/GO-102009-2913). Office of Energy Efficiency and Renewable Energy.
9. European Industrial Gases Association AISBL. 2024. Standard Procedures for Hydrogen Supply Systems. Doc 250/24. Brussels: EIGA.
10. Sand, M., Fuglestvedt, J.S., Lund, M.T., Samset, B.H. and Shine, K.P. 2023. A multi-model assessment of the global warming potential of hydrogen. Commun. Earth Environ. 4(1): 203.
11. Smith, N.J.P. 2002. It's time for explorationists to take hydrogen more seriously. First Break 20: 246–253.
12. Maiga, O., Deville, E., Laval, J., Prinzhofer, A. and Diallo, A.B. 2023. Characterization of the spontaneously recharging natural hydrogen reservoirs of Bourakebougou in Mali. Sci. Rep. 13: 11876.
13. Zgonnik, V. 2020. Occurrence and geoscience of natural hydrogen. Int. J. Hydrogen Energy 45(39): 21039–21061.
14. The Economist. 2023. Meet the boffins and buccaneers drilling for hydrogen. 23 Dec: 65–68.
15. Ellis, G.S. and Gelman, S.E. 2024. Model predictions of global geologic hydrogen resources. Sci. Adv. 10(50).
16. Singh, H. 2025. The emergence of natural hydrogen: Genesis and current perspectives. Scientific Contributions Oil and Gas 48(1): 91–112.
17. International Energy Agency. 2023. Global Hydrogen Review 2023. IEA, Paris.
18. International Energy Agency. 2024. World Energy Outlook 2024. IEA, Paris.
19. IRENA. 2022. Geopolitics of the Energy Transformation: The Hydrogen Factor. Abu Dhabi: International Renewable Energy Agency.

Chapter 2

1. International Energy Agency. 2023. Global Hydrogen Review 2023, p. 77. IEA, Paris.
2. Avargani, V., Rahbari, A., Alizadehdakhel, A. and Hanafizadeh, P. 2022. A comprehensive review on hydrogen production and utilization in North America: Prospects and challenges. Energy Convers. Manag. 269: 116086.
3. International Energy Agency. 2022. Global Hydrogen Review 2022. IEA, Paris.
4. Mengdi, J. and Wang, J. 2021. Review and comparison of various hydrogen production methods based on costs and life cycle impact assessment indicators. Int. J. Hydrogen Energy 46: 38612–38635.
5. International Energy Agency. 2020. CCUS in Clean Energy Transitions. IEA, Paris.
6. Patlolla, A., Subramaniam, M. and Ozcan, S. 2023. A review of methane pyrolysis technologies for hydrogen production. Renew. Sustain. Energy Rev. 181: 113371.
7. Harrison, S.B. 2021. Turquoise hydrogen production by methane pyrolysis. Pet. Technol. Q. Q4: 73–76.
8. Fulcheri, L., Rohani, V.-J., Wyse, E., Hardman, N. and Dames, E. 2023. An energy efficient plasma methane pyrolysis process for high yields of carbon black and hydrogen. Int. J. Hydrogen Energy 48(70): 26491–26502.
9. International Energy Agency. 2023. Global Hydrogen Review 2023, pp. 87–88. IEA, Paris.
10. Zero Carbon Analytics. 2023. A closer look at CCS: Problems and potential. Retrieved April 22, 2025.
11. Robertson, B. 2022. The Carbon Capture Crux: Lessons Learned. Institute for Energy Economics and Financial Analysis (IEEFA).
12. Abdulla, A., Hanna, R., Schell, K.R., Babacan, O. and Victor, D.G. 2021. Explaining successful and failed investments in U.S. carbon capture and storage using empirical and expert assessments. Environ. Res. Lett. 16(1): 014036.
13. Kearns, D., Liu, H. and Consoli, C. 2021. Technology Readiness and Costs of CCS. Global CCS Institute.
14. Global CCS Institute. 2021. Technology Readiness and Costs of CCS. Melbourne: Global CCS Institute.
15. International Energy Agency. 2021. Is Carbon Capture too Expensive? IEA, Paris.
16. Hauber, G. 2023. Norway's Sleipner and Snøhvit CCS: Industry Models or Cautionary Tales? Institute for Energy Economics and Financial Analysis.
17. World Bank. 2024. State and Trends of Carbon Pricing 2024. Washington, DC: World Bank.
18. Jones, A.C. and Lawson, A.J. 2022. Carbon Capture and Sequestration in the United States (CRS Report No. R44902), pp. 16–17. Congressional Research Service.
19. Hauber, G. 2023. Norway's Sleipner and Snøhvit CCS: Industry models or cautionary tales? Institute for Energy Economics and Financial Analysis, 14 June.
20. Rubin, E.S., Davison, J.E. and Herzog, H.J. 2015. The cost of CO_2 capture and storage. Int. J. Greenhouse Gas Control 40: 378–400.
21. International Energy Agency. 2020. Projected Costs of Generating Electricity 2020. IEA, Paris.
22. Hydrogen Council. 2021. Hydrogen Decarbonisation Pathways: A Life-Cycle Assessment.
23. Sun, T., Shrestha, E., Hamburg, S.P., Kupers, R. and Ocko, I.B. 2024. Climate impacts of hydrogen and methane emissions can considerably reduce the climate benefits across key hydrogen use cases and time scales. Environ. Sci. Technol. 58(12): 5299–5309.
24. Howarth, R.W. and Jacobson, M.Z. 2021. How green is blue hydrogen? Energy Sci. Eng. 9(10): 1676–1687.
25. UNESCO. 2024. United Nations World Water Development Report 2024: Water for Prosperity and Peace, p. 14. United Nations Educational, Scientific and Cultural Organization (UNESCO).
26. Baykara, S.Z. 2018. Hydrogen: A brief overview on its sources, production and environmental impact. Int. J. Hydrogen Energy 43(23): 10605–10614.
27. International Energy Agency. 2021. Methane Tracker 2021. IEA, Paris.
28. International Energy Agency. 2024. Global Hydrogen Review 2024. IEA, Paris.

29. Sánchez-Bastardo, N., Barrios, E. and Röntzsch, L. 2021. Methane pyrolysis for zero-emission hydrogen production: A potential bridge technology from fossil fuels to a renewable and sustainable hydrogen economy. Int. J. Hydrogen Energy 46(33): 17001–17022.
30. Lazard. 2024. Levelized Cost of Energy+, Levelized Cost of Storage, and Levelized Cost of Hydrogen – Version 17.0. Lazard Ltd.
31. Robertson, B. and Mousavian, M. 2022. Reality Check on CO_2 Emissions Capture at Hydrogen-From-Gas Plants. Institute for Energy Economics and Financial Analysis (IEEFA).
32. Liu, H., Zhang, X., Zhai, H. and Xu, H. 2022. Carbon footprint analysis of coal-based hydrogen production with CCS technologies: A case study in China. J. Clean. Prod. 366: 132973.

Chapter 3

1. International Energy Agency (IEA). 2023. Energy Technology Perspectives 2023. Paris: IEA.
2. International Gas Union (IGU). 2024. World LNG Report – 2024 Edition. Barcelona, Spain: IGU.
3. International Renewable Energy Agency (IRENA). 2022. Global Hydrogen Trade to Meet the 1.5°C Climate Goal: Part I – Trade Outlook for 2050 and Way Forward. Abu Dhabi: IRENA.
4. Gunkel, P.A., Schneider, L. and Schill, W.-P. 2022. Hydrogen road transport analysis in the energy system: A case study for Germany through 2050. Energy Rep. 8: 3437–3453.
5. Energy Transitions Commission (ETC). 2021. Making the Hydrogen Economy Possible: Accelerating Clean Hydrogen in an Electrified Economy. London: Energy Transitions Commission, p. 38.
6. Steiner, M., Marewski, U. and Silcher, H. 2023. Investigation of Steel Materials for Gas Pipelines and Plants for Assessment of Their Suitability with Hydrogen: Final Report of the DVGW Research Project SyWeSt H$_2$ (G 202006). Bonn: Deutscher Verein des Gas- und Wasserfaches (DVGW).
7. Topolski, K., Reznicek, E.P., Erdener, B.C., San Marchi, C.W., Ronevich, J.A., Fring, L. et al. 2022. Hydrogen Blending into Natural Gas Pipeline Infrastructure: Review of the State of Technology. National Renewable Energy Laboratory (NREL), Golden, CO. NREL/TP-5400-81704.
8. Energy Transitions Commission (ETC). 2021. Making the Hydrogen Economy Possible: Accelerating Clean Hydrogen in an Electrified Economy. London: ETC, p. 38.
9. Esquivel-Elizondo, S., Hormaza Mejia, A., Sun, T., Shrestha, E., Hamburg, S.P. and Ocko, I.B. 2023. Wide range in estimates of hydrogen emissions from infrastructure. Front. Energy Res. 11: 1207208.
10. Ogge, M. 2022. Brown Coal, Greenwash: The True Emissions Impact of the Hydrogen Energy Supply Chain Project. The Australia Institute.
11. DeSantis, D., James, B.D., Houchins, C., Saur, G. and Lyubovsky, M. 2021. Cost of long-distance energy transmission by different carriers. iScience 24(12): 103495.
12. Yang, M., Hunger, R., Berrettoni, S., Sprecher, B. and Wang, B. 2023. A review of hydrogen storage and transport technologies. Clean Energy 7(1): 190–216.
13. International Renewable Energy Agency (IRENA). 2022. Global Hydrogen Trade to Meet the 1.5°C Climate Goal: Part II – Technology Review of Hydrogen Carriers. Abu Dhabi: IRENA, p. 131.

Chapter 4

1. Latenser, B.A. and Lucktong, K. 2000. Anhydrous ammonia burns: Case presentation and literature review. Burns 26(4): 407–410.
2. International Renewable Energy Agency (IRENA). 2022. Global Hydrogen Trade to Meet the 1.5°C Climate Goal: Part II – Technology Review of Hydrogen Carriers. Abu Dhabi: IRENA.
3. Al-Beiker, A.R. and Farooq, S. 2023. Comparative cost assessment of sustainable energy carriers produced from natural gas accounting for boil-off gas and social cost of carbon. Int. J. Hydrogen Energy 48(34): 12773–12790.
4. Kang, K., Kim, J. and Yoon, Y. 2022. A review on ammonia blends combustion for industrial applications. Fuel 316: 123346.

5. Rouwenhorst, K.H.R., van der Ham, L.V., Mul, G. and Kersten, S.R.A. 2019. Islanded ammonia power systems: Technology review and conceptual process design. Renew. Sustain. Energy Rev. 114: 109339.

6. Huangang, T., Xinyue, Z., Yixuan, Z., Ziyan, C., Cheng, T. and Yang, G. 2024. Advances in power generation from ammonia via electrocatalytic oxidation in direct ammonia fuel cells. J. Power Sources 600: 233478.

7. International Energy Agency (IEA). 2023. Unlocking the Potential of Direct Air Capture: Is Scaling Up Through Carbon Markets Possible? IEA, Paris.

8. International Energy Agency (IEA). 2024. The Role of E-fuels in Decarbonising Transport. Paris: IEA.

9. Ueckerdt, F. 2023. E-Fuels – Aktueller Stand und Projektionen. Potsdam Institute for Climate Impact Research (PIK).

10. Stratton, R.W., Wong, H.M. and Hileman, J.I. 2010. Life cycle greenhouse gas emissions from alternative jet fuels. Partnership for Air Transportation Noise and Emissions Reduction (PARTNER), Project 28 Report.

11. International Energy Agency (IEA). 2023. Renewables 2023: Transport Biofuels. Paris: IEA.

12. IEA Bioenergy. 2024. Sustainable Aviation Fuels: Status, Barriers and Opportunities. Paris: International Energy Agency.

13. International Energy Agency (IEA). 2021. Fossil Jet and Biojet Fuel Production Cost Ranges 2010–2030. Paris: IEA.

14. Xu, H., Ou, L., Li, Y., Hawkins, T.R. and Wang, M. 2022. Life cycle greenhouse gas emissions of biodiesel and renewable diesel production in the United States. Environ. Sci. Technol. 56(12): 7512–7521.

Chapter 5

1. International Energy Agency (IEA). 2023. Global Hydrogen Review 2023. Paris: IEA.

2. International Energy Agency (IEA). 2021. Ammonia Technology Roadmap: Towards More Sustainable Nitrogen Fertiliser Production. Paris: IEA.

3. The Royal Society. 2020. Green Ammonia: Policy Briefing. London: The Royal Society.

4. Saygin, D., Hammingh, P. and Diermann, L. 2023. Ammonia production from clean hydrogen and the implications for global natural gas demand. Utrecht University, Copernicus Institute of Sustainable Development, Utrecht.

5. International Energy Agency (IEA). 2023. Energy Technology Perspectives 2023. Paris: IEA.

6. International Energy Agency (IEA). 2023. Tracking Clean Energy Progress 2023. Paris: IEA.

7. International Renewable Energy Agency (IRENA) and Methanol Institute. 2021. Innovation Outlook: Renewable Methanol. Abu Dhabi: IRENA.

8. International Energy Agency (IEA). 2024. Global Hydrogen Review 2024. Paris: IEA.

9. Methanol Institute. 2022. The Carbon Footprint of Methanol. Washington, DC: Methanol Institute.

10. International Energy Agency (IEA). 2021. Solar Energy Policy in Uzbekistan: A Roadmap. Paris: IEA.

11. Deloitte, World Wildlife Fund. 2023. Assessment of Green Hydrogen for Industrial Heat. Renewable Thermal Collaborative.

12. CEMBUREAU. 2020. Cementing the European Green Deal: The European Cement Industry's Ambition for 2050. Brussels: The European Cement Association.

13. World Steel Association. 2024. World Steel in Figures 2024. Brussels, Belgium: World Steel Association.

14. Carey, L., Brickman, A., Yavorsky, N., Gamage, C. and Rosas, J. 2023. Opportunities for Near-Zero-Emissions Steel Production in the Great Lakes. Rocky Mountain Institute.

15. Institute for Energy Economics and Financial Analysis (IEEFA). 2022. Steel Sector – Facts About Emissions and Decarbonisation Technologies.
16. World Steel Association. n.d. Fact Sheet: Energy Use in the Steel Industry. Brussels, Belgium: World Steel Association.
17. Deloitte. 2023. Green Steel: The Race to Net-Zero Construction. Deloitte Touche Tohmatsu Limited.

Chapter 6

1. International Energy Agency (IEA). 2024. Renewables 2023. Paris: IEA.
2. UK Department for Energy Security and Net Zero (DESNZ). 2023. Hydrogen Heating Overview.
3. International Energy Agency (IEA). 2022. The Future of Heat Pumps. Paris: IEA.
4. Rosnow, J. 2024. Hydrogen for heating? A meta-review of the evidence. Energy Policy Stud. 38(2): 112–131.
5. Fraunhofer Institute for Solar Energy Systems ISE. 2020. Hydrogen in the Energy System of the Future: Focus on Heat in Buildings, p. 5. Freiburg, Germany.
6. Arup. 2023. The Future of Great Britain's Gas Networks. Report prepared for the National Infrastructure Commission.
7. Electricity System Operator (ESO). 2024. Beyond 2030: A National Blueprint for a Decarbonised Electricity System in Great Britain. National Grid Electricity System Operator.
8. Scholand, M. 2024. Debunking the Hydrogen Hype: Why Europe Should Pursue Electrification Instead of Hydrogen for Cooking and Heating. Environmental Coalition on Standards (ECOS).
9. ARUP. 2021. Hy4Heat Safety Assessment Conclusions Report. Prepared for the UK Department for Business, Energy and Industrial Strategy (BEIS).
10. Frazer-Nash Consultancy. 2018. Appraisal of Domestic Hydrogen Appliances. FNC 55089/46433R Issue 1, p. 18.
11. Wright and Lewis. 2022. Emissions of NOx from blending of hydrogen and natural gas in space heating boilers.
12. National Audit Office. 2024. Decarbonising Home Heating. UK Government.
13. Competition and Markets Authority (CMA). 2023. Consumer Protection in the Green Heating and Insulation Sector. 14 December 2023.
14. International Energy Agency (IEA). 2023. Global Installed Energy Storage Capacity by Scenario, 2023 and 2030. Paris: IEA.
15. U.S. Department of Energy. 2022. Grid Energy Storage Technology Cost and Performance Assessment. Office of Electricity, U.S. Department of Energy, September 2022.
16. Pashchenko, D. 2024. Ammonia fired gas turbines: Recent advances and future perspectives. Energy 290: 130275.
17. Cesaro, Z., Ives, M., Nayak-Luke, R., Mason, M. and Bañares-Alcántara, R. 2021. Ammonia to power: Forecasting the levelized cost of electricity from green ammonia in large-scale power plants. Appl. Energy 282: 116009.
18. BloombergNEF. 2022. Japan's Costly Ammonia Coal Co-Firing Strategy: A Costly Approach to Decarbonization—Renewables Present More Economic Alternative. 28 September 2022.
19. Ministry of the Environment, Japan. 2023. Communiqué of the G7 Ministers' Meeting on Climate, Energy and Environment. April 2023.

Chapter 7

1. International Energy Agency. 2021. Well-to-wheel (wake/wing) GHG intensity of motorised passenger transport modes, IEA, Paris https://www.iea.org/data-and-statistics/charts/well-to-wheel-wake-wing-ghg-intensity-of-motorised-passenger-transport-modes, Licence: CC BY 4.0.
2. International Energy Agency (IEA). 2019. The Future of Rail. Paris: IEA.

3. Liu, X., Reddi, K., Elgowainy, A., Lohse-Busch, H., Wang, M. and Rustagi, N. 2020. Comparison of well-to-wheels energy use and emissions of a hydrogen fuel cell electric vehicle relative to a conventional gasoline-powered internal combustion engine vehicle. Int. J. Hydrogen Energy 45(1): 972–983.
4. Yoo, E., Kim, M. and Song, H.H. 2018. Well-to-wheel analysis of hydrogen fuel-cell electric vehicle in Korea. Int. J. Hydrogen Energy 43(41): 19267–19278.
5. Bossel, U. 2006. Does a hydrogen economy make sense? Proc. IEEE 94(10): 1826–1837.
6. International Energy Agency (IEA). 2021. Well-to-Wheels Greenhouse Gas Emissions for Cars by Powertrains. Paris: IEA.
7. Woo, J.R., Choi, H. and Ahn, J. 2017. Well-to-wheel analysis of greenhouse gas emissions for electric vehicles based on electricity generation mix: A global perspective. Transp. Res. Part D Transp. Environ. 51: 340–350.
8. International Energy Agency (IEA). 2024. Global EV Outlook 2024: Trends and Developments in Electric Vehicle Markets, Charging Infrastructure and Battery Demand. Paris: IEA.
9. Correa, G., Muñoz, P. and Poggio, A. 2019. A comparative energy and environmental analysis of a diesel, hybrid, hydrogen and electric urban bus. Energy 187: 115906.
10. DNV. 2023. Hydrogen Forecast to 2050. Høvik, Norway: DNV AS.
11. International Energy Agency (IEA). 2024. Global EV Data Explorer. Paris: IEA.
12. Association of American Railroads. 2021. Oppose Rail Electrification & Support Sensible Climate Policy. February [Fact Sheet].
13. Hoffrichter, A., Miller, A.R., Hillmansen, S. and Roberts, C. 2012. Well-to-wheel analysis for electric, diesel and hydrogen traction for railways. Transp. Res. Part D Transp. Environ. 17(1): 28–34.
14. Prussi, M., Yugo, M., De Prada, L., Padella, M. and Edwards, R. 2020. JEC Well-to-Wheels Report v5 (EUR 30284 EN). Publications Office of the European Union.
15. Klöwer, M., Allen, M.R., Lee, D.S., Proud, S.R., Gallagher, L. and Skowron, A. 2021. Quantifying aviation's contribution to global warming. Environ. Res. Lett. 16(10): 104027.
16. Schäfer, A.W., Barrett, S.R.H., Doyme, K., Dray, L.M., Gnadt, A.R., Self, R. et al. 2019. Technological, economic and environmental prospects of all-electric aircraft. Nat. Energy 4(2): 160–166.
17. Mukhopadhaya, J. and Rutherford, D. 2022. Performance Analysis of Evolutionary Hydrogen-Powered Aircraft. White Paper. International Council on Clean Transportation.
18. International Air Transport Association. 2023. Energy and New Fuels Infrastructure: Net Zero Roadmap.
19. Stratton, R.W., Wong, H.M. and Hileman, J.I. 2010. Life cycle greenhouse gas emissions from alternative jet fuels. Partnership for Air Transportation Noise and Emissions Reduction (PARTNER), Project 28 Report.
20. IEA Bioenergy. 2024. Sustainable Aviation Fuel: Review of Production Pathways and Status in Participating Countries. IEA Bioenergy Task 39, June.
21. Lee, D.S., Fahey, D.W., Skowron, A., Allen, M.R., Burkhardt, U., Chen, Q. et al. 2021. The contribution of global aviation to anthropogenic climate forcing for 2000 to 2018. Atmos. Environ. 244: 117834.
22. Guyon, O., Lucas, M., Maricar-Pichon, M., El-Kadi, J. and Dauphin, R. 2025. Life cycle assessment of e-/bio-methanol and e-/grey-/blue-ammonia for maritime transport. IFP Energ. Nouv.
23. Lee, H., Lee, J., Roh, G., Lee, S., Choung, C. and Kang, H. 2024. Comparative life cycle assessments and economic analyses of alternative marine fuels: Insights for practical strategies. Sustainability 16(5): 2114.
24. International Energy Agency (IEA). 2023. The Role of E-fuels in Decarbonising Transport. Paris: IEA.
25. American Bureau of Shipping (ABS). 2021. Methanol as Marine Fuel. Houston, TX: ABS.
26. Dawson, L., Ware, J. and Vest, L. 2022. Ammonia at Sea: Studying the Potential Impact of Ammonia as a Shipping Fuel on Marine Ecosystems. Environmental Defense Fund.

27. International Energy Agency (IEA). 2023. Renewables 2023: Analysis and Forecast to 2028. Paris: IEA.
28. Verschuur, J., Salmon, N., Hall, J. and Bañares-Alcántara, R. 2024. Optimal fuel supply of green ammonia to decarbonise global shipping. Environ. Res. Infrastruct. Sustain. 4(1): 015001.

Chapter 8

1. International Renewable Energy Agency (IRENA). 2020. Green Hydrogen Cost Reduction: Scaling up Electrolysers to Meet the 1.5°C Climate Goal. Abu Dhabi: IRENA.
2. International Solar Alliance. 2023. World Solar Technology Report 2023. Gurgaon, India.
3. Nøland, J.K., Auxepaules, J., Rousset, A., Perney, B. and Falletti, G. 2022. Spatial energy density of large-scale electricity generation from power sources worldwide. Scientific Data 9: 709.
4. Imam, A.A., Abusorrah, A. and Marzband, M. 2024. Potentials and opportunities of solar PV and wind energy sources in Saudi Arabia: Land suitability, techno-socio-economic feasibility, and future variability. Results in Engineering 21: 101785.
5. International Energy Agency (IEA). 2024. World Energy Outlook 2024. Paris: IEA.
6. Tonelli, D., Rosa, L., Gabrielli, P., Caldeira, K., Parente, A. and Contino, F. 2023. Global land and water limits to electrolytic hydrogen production using wind and solar resources. Nature Communications 14: 5532.
7. Hydrogen Council and McKinsey & Company. 2024. Hydrogen Insights 2024. Hydrogen Council.

Chapter 9

1. International Energy Agency (IEA). 2024. Global Hydrogen Review 2024, Paris: IEA.
2. United Nations Industrial Development Organization (UNIDO), International Renewable Energy Agency (IRENA), and German Institute of Development and Sustainability (IDOS). 2023. Reframing the Narrative on Green Hydrogen: A Policy Toolkit for Developing Countries. Vienna: UNIDO.
3. Ritchie, H. 2024. Solar panel prices have fallen by around 20% every time global capacity doubled. Our World in Data, June.
4. International Energy Agency (IEA). 2024. Strategies for Affordable and Fair Clean Energy Transitions. Paris: IEA.
5. BloombergNEF (BNEF). 2024. Energy Transition Factbook 2024. London: Bloomberg Finance L.P.

Chapter 10

1. Eurostat. 2024. Energy Statistics—An Overview. Luxembourg: Publications Office of the European Union.
2. Adams, M., Nyathi, J., Sood, S. and Davie, A. 2023. Weighing the EU Options: Importing Versus Domestic Production of Hydrogen & E-Fuels. Report prepared by Ricardo Energy & Environment for Transport & Environment (T&E), p. 5.
3. Agency for the Cooperation of Energy Regulators (ACER). 2024. Hydrogen Market Monitoring Report 2024. Ljubljana: ACER.
4. Hydrogen Europe. 2025. Hydrogen Europe Quarterly, Issue 10 (Q1 2025).
5. Cheng, C., Blakers, A., Stocks, M. and Lu, B. 2022. 100% Renewable Energy in Japan. Energy Conversion and Management 263: 115677.
6. Park, W.Y., Fridley, D., Ahn, H., Kim, J. and Greenblatt, J. 2023. A Clean Energy Korea by 2035: Transitioning to 80% Carbon-Free Electricity Generation. Lawrence Berkeley National Laboratory, Report No. LBNL-2001523.
7. Ulreich, S. 2024. Turning Hydrogen Demand Into Reality: Which Sectors Come First? International Chamber of Shipping.
8. Singh, R. 2024. Korea's low-carbon hydrogen power tenders may need sweeteners to succeed: experts. S&P Global Commodity Insights, 16 December 2024 [Web Edition].

9. U.S. Department of Energy (DOE). 2024. Pathways to Commercial Liftoff: Clean Hydrogen. Washington, D.C.: DOE.
10. International Energy Agency (IEA). 2024. Electricity 2024. Paris: IEA.
11. International Energy Agency (IEA). 2025. Gas Market Report, Q1-2025. Paris: IEA.

Chapter 11

1. Campbell, R. and Ryan, M. 2024. Hope and Hydrogen – Australia's Hydrogen Export Charade. The Australia Institute.
2. International Energy Agency (IEA). 2023. Renewable Hydrogen from Oman: A Producer Economy in Transition. Paris: IEA.
3. International Renewable Energy Agency (IRENA). 2022. Geopolitics of the Energy Transformation: The Hydrogen Factor, p. 53. Abu Dhabi: IRENA.
4. Adams, M., Nyathi, J., Sood, S. and Davie, A. 2023. Weighing the EU Options: Importing Versus Domestic Production of Hydrogen & E-Fuels. Report prepared by Ricardo Energy & Environment for Transport & Environment (T&E).
5. Scholten, D., Westphal, K., Kemmerzell, J. and Gandenberger, C. 2023. Global Hydrogen Economy – Hype or Hope? Opportunities and Risks for Partner Countries in the Global South. Bonn: German Institute of Development and Sustainability (IDOS).
6. Energy Industries Council (EIC). 2025. Africa OPEX Report 2025.
7. International Energy Agency (IEA). 2023. Latin America Energy Outlook 2023. Paris: IEA.

Chapter 12

1. U.S. Department of Energy (DOE). 2023. Pathways to Commercial Liftoff: Clean Hydrogen. Washington, D.C.: U.S. Department of Energy, March.
2. International Energy Agency (IEA). 2024. Global Hydrogen Review 2024, p. 21. Paris: IEA.
3. Xiang, P.-P., Zhang, H., Wang, W., Yu, Y., Xu, Y., Wang, W. et al. 2023. Role of hydrogen in China's energy transition towards carbon neutrality target: IPAC analysis. Adv. Clim. Change Res. 14(1): 43–48.
4. Hydrogen Energy Circle. 2019. China Hydrogen Energy and Fuel Cell Investment Project Inventory. Summary provided via translation by Feng Jiang (Beijing Normal University), personal communication, 2023.
5. Tao, W. 2021. Hydrogen Development in China: A Policy Perspective. Oxford Energy Forum 127 (May): 36–41. Oxford Institute for Energy Studies.
6. New Energy Vehicle National Big Data Alliance. 2020. National Hydrogen Fuel Cell Vehicle Data Analysis Report. Summary provided via translation by Feng Jiang (Beijing Normal University), personal communication, 2023.
7. Liu, W., Wan, Y., Xiong, Y. and Gao, P. 2021. Green hydrogen standard in China: Standard and evaluation of low-carbon hydrogen, clean hydrogen, and renewable hydrogen. pp. 211–224. In: Li, Y., Phoumin, H. and Kimura, S. (eds.). Hydrogen Sourced from Renewables and Clean Energy: A Feasibility Study of Achieving Large-scale Demonstration. Jakarta: ERIA. See p. 217.
8. National Development and Reform Commission (NDRC). 2024. Guiding Opinions on Vigorously Implementing Renewable Energy Substitution Actions (NDRC Energy [2024] No. 1537). [In Chinese].
9. Climate Change Committee. 2025. The Seventh Carbon Budget: The UK's Path to Net Zero, p. 239. London: CCC.
10. Duckett, A. 2024. UK invests £22bn in CCS clusters and hydrogen projects. The Chem. Eng. 21 March 2024 [web edition].
11. International Energy Agency (IEA). 2021. India Energy Outlook 2021, p. 71. Paris: IEA.

12. Ministry of New and Renewable Energy (MNRE). 2023. National Green Hydrogen Mission: Strategy Paper, p. 2. Government of India, New Delhi.
13. International Energy Agency (IEA). https://www.iea.org/countries/india (accessed March 2024).
14. Ministry of Coal. 2021. National Coal Gasification Mission: Mission Document. Government of India, New Delhi.

Phantom or Panacea?

1. Georgescu-Roegen, N. 1984. Feasible recipes versus viable technologies. Atl. Econ. J. 12(1): 21–31.

Index

Endorsements

Over the last two centuries, fossil fuels have supported the growth of world economies, helped by our lack of awareness of the environmental consequences of CO_2 emissions. Recent worldwide climate change impacts are pushing researchers and policy makers towards new and less damaging energy sources. Glucina and Mayumi explore one of the most attractive alternatives: the "hydrogen economy". Hydrogen, the smallest but also the most abundant atomic species in the biosphere, has a very special characteristic. Burning it does not release CO_2, simply water vapor, which does not contribute to global warming. Very small amounts of pure hydrogen are available in the biosphere, although it is abundant in the form of water, organic matter, and minerals (not to mention hydrocarbons). The most promising clean production method, electrolysis, uses electricity from renewable sources to split water into "green" hydrogen and oxygen. Hydrogen, stored in liquid form, can be used to fuel vehicles, support industrial chemistry, and generate electricity and heat – even when renewable sources are not available, like during the night. Is hydrogen a magic bullet to solve societal energy problems? Aware of the challenges, the authors clear a path to the decarbonized era. A fascinating journey through technologies, cost barriers, and opportunities, on route to a sustainable low-carbon future.

Sergio Ulgiati

Professor of Environmental Chemistry
Department of Science and Technology
Parthenope University of Napoli, Italy

Hydrogen Energy in a Sustainable Future: Phantom or Panacea? offers a critical and timely analysis of hydrogen's potential and limitations as a clean energy carrier. While often promoted as the future of energy, hydrogen faces fundamental challenges, including low volumetric energy density, costly storage, and heavy reliance on government subsidies. The book provides a realistic perspective, highlighting why hydrogen is unlikely to scale quickly enough to meet early decarbonization goals.

This aligns with post-normal science (PNS), which addresses complex, high-stakes issues with deep uncertainties and political influences. Hydrogen's future is not just a technical challenge but a socio-political one, shaped by policy shifts, economic constraints, and technological hurdles. The book exemplifies how

PNS can guide energy policy by questioning dominant narratives and promoting adaptive, multi-stakeholder approaches. By critically evaluating hydrogen's role in decarbonization, this book by ***Mark Glucina and Kozo Mayumi*** contributes to a more informed and balanced discussion on sustainable energy transitions.

Silvio Funtowicz

European Centre for Governance in Complexity – (Nesttun, Norway)
(https://www.ecgc.no/)
Centre for the Study of the Sciences and the Humanities – University of
Bergen (Norway) (https://uib.no/en/svt)

For Product Safety Concerns and Information please contact our EU
representative GPSR@taylorandfrancis.com
Taylor & Francis Verlag GmbH, Kaufingerstraße 24, 80331 München, Germany

www.ingramcontent.com/pod-product-compliance
Lightning Source LLC
Chambersburg PA
CBHW060552220326
41598CB00024B/3084